Springer Tracts in Natural Philosophy

Volume 16

Edited by B. D. Coleman

Co-Editors: R. Aris · L. Collatz · J. L. Ericksen
P. Germain · M. E. Gurtin · M. M. Schiffer
E. Sternberg · C. Truesdell

Paul Slepian

Mathematical Foundations
of Network Analysis

Springer-Verlag New York Inc. 1968

Paul Slepian
Department of Mathematics
Rensselaer Polytechnic Institute
Troy, New York

ISBN-13: 978-3-642-87426-0   e-ISBN-13: 978-3-642-87424-6
DOI: 10.1007/978-3-642-87424-6

© by Springer-Verlag Berlin · Heidelberg 1968

Softcover reprint of the hardcover 1st edition 1968

Library of Congress Catalog Card Number 68-29708.

Title No. 6744

To Laura and Jean

Acknowledgements

A small part of this book was written during the summer of 1966, when the author was employed in the Applied Mathematics Division of Argonne National Laboratory. The author would like to express his appreciation to the Applied Mathematics Division and the United States Atomic Energy Commission's Argonne National Laboratory for their generous support during this period.

The author would also like to thank Mr. Timothy A. Loughlin, who carefully read the entire manuscript and suggested many improvements.

Contents

Chapter 6. Axioms of Network Analysis

Chapter 7. Existence and Uniqueness of Solutions

Chapter 8. Kirchhoff's Third and Fourth Laws

Introduction

In this book we attempt to develop the fundamental results of resistive network analysis, based upon a sound mathematical structure. The axioms upon which our development is based are Ohm's Law, Kirchhoff's Voltage Law, and Kirchhoff's Current Law. In order to state these axioms precisely, and use them in the development of our network analysis, an elaborate mathematical structure is introduced, involving concepts of graph theory, linear algebra, and one dimensional algebraic topology.

The graph theory and one dimensional algebraic topology used are developed from first principles; the reader needs no background in these subjects. However, we do assume that the reader has some familiarity with elementary linear algebra. It is now stylish to teach elementary linear algebra at the sophomore college level, and we feel that the requirement that the reader should be familiar with elementary linear algebra is no more demanding than the usual requirement in most electrical engineering texts that the reader should be familiar with calculus. In this book, however, no calculus is needed. Although no formal training in circuit theory is needed for an understanding of the book, such experience would certainly help the reader by presenting him with familiar examples relevant to the mathematical abstractions introduced.

It is our intention in this book to exhibit the effect of the topological properties of the network upon the branch voltages and branch currents, the objects of interest in network analysis. This effect is most easily exhibited when the branch elements of the network have the simplest electrical character. Consideration of more complicated branch elements would lead to more complicated results which would obscure the relation between the topology and network analysis that we are trying to expose. Thus, we limit ourselves to resistive networks in this book. This suggests, of course, that an area of future work in this direction is an extension of the results of this book to networks with more complicated branch elements.

Another fertile area for the extension of the results of this book is the field of network synthesis. Indeed, the writer was led to the production of this book by years of work in network synthesis, during which he was severely limited by the lack of precision of the network analysis. It is hoped that with the more precise tools of network analysis developed in this book, certain areas of network synthesis, previously obscure, may be clarified.

This book is not intended as a text for a formal course; no exercises are included. However, it may be suitable for seminar work. Hopefully, such a seminar would be populated by mathematically sophisticated electrical engineers desiring a more precise exposition of network analysis, and by mathematicians curious to learn how the abstractions of topology can be applied to the realities of electrical network theory.

CHAPTER ONE

Connected Networks

1.0 Introduction

In this chapter we introduce networks and investigate their simplest geometrical properties. Our networks for the present will be divorced from all circuit theory considerations. Such a divorce allows us to examine those network properties which are independent of circuit theory considerations. Much later in the book, after our elementary machinery is firmly established, we shall introduce the complications of circuit theory into our existing elementary structures.

1.1 Set Theory

In order to achieve precision in our exposition, we shall use the language of elementary set theory. The reader should bear in mind that our set theory is merely a notational language, and we use this language in its most simple form, coinciding with level of set theory taught in most high-schools at present. Most readers are probably familiar with the simple set theory which we shall use. However, for the sake of completeness, we shall exhibit here the notations and definitions to be used.

In particular, we adopt the naive point of view that every object in the world is a set, and if x and B are sets, we write

$$x \in B$$

as an abbreviation for the statement "x is an element of B," while we write

$$x \notin B$$

as an abbreviation for the statement "x is not an elment of B."

Vital to our logic is the assumption that for any x and any B, the statement $x \in B$ is a proposition, and must be true or false.

If for each set x, $p(x)$ is a proposition, then we write

$$\{x \mid p(x)\}$$

as an abbreviation for the set of all x such that $p(x)$ is true.

We define

$$A \text{ is a subset of } B$$
$$\text{if and only if}$$
$$\text{for all } x, \; x \in A \text{ implies } x \in B,$$

and we write

$$A \subset B$$

as an abbreviation for the statement "A is a subset of B."

As expected, we define

$$A = B$$
$$\text{if and only if}$$
$$A \subset B \text{ and } B \subset A,$$

and we write $A \neq B$ if and only if the proposition $A = B$ is false.

Finally, we define the intersection, union, and difference of two sets in the usual way. In particular,

$$A \cap B = \{x \mid x \in A \text{ and } x \in B\}.$$
$$A \cup B = \{x \mid x \in A \text{ or } x \in B\}.$$
$$A - B = \{x \mid x \in A \text{ and } x \notin B\}.$$

1.2 Sets with Two or Less Elements

We shall use the symbol 0 for the empty set as well as the number zero. Formally, we define the empty set by

$$0 = \{x \mid x \neq x\},$$

and note that the statement $0 \subset B$ is true for every set B.

For any set x we shall let $\{x\}$ be the set whose only element is x. Formally,

$$\{x\} = \{y \mid y = x\}.$$

Finally, we define

$$\{x, y\} = \{x\} \cup \{y\},$$

and observe that $\{x, y\}$ has two elements when $x \neq y$, but only one element when $x = y$.

1.3 Generalized Union

For a set B we shall frequently want to consider the union of all elements of B. We denote this union by σB, and define formally,

$$\sigma B = \{x \mid \text{for some } y,\ x \in y \text{ and } y \in B\}.$$

Our notation σB is not too common. The reader may have seen σB described in the literature by $\bigcup_{x \in B} x$ or perhaps $\bigcup B$.

1.4 Relations and Functions

In keeping with modern usage we shall consistently treat functions as sets of ordered pairs with an additional property to be specified.

In particular, for any x and y we write the ordered pair (x, y) in the usual way and say that x is the first coordinate of the ordered pair (x, y) and y is the second coordinate of the ordered pair (x, y). Observe that we do not consider the elements of the ordered pair (x, y). Such a consideration of the elements of an ordered pair requires a more sophisticated treatment of set theory.

We say that two ordered pairs are equal if and only if their respective coordinates are equal. Formally,

$$(x, y) = (u, v)$$
if and only if
$$x = u \text{ and } y = v.$$

We must distinguish carefully between the ordered pair (x, y) and the set $\{x, y\}$. For example

$$\{x, y\} = \{u, v\}$$
if and only if
$$x = u \text{ and } y = v, \text{ or, } x = v \text{ and } y = u.$$

A relation is a set of ordered pairs. A function is a relation such that distinct ordered pairs of the relation must have different first coordinates. Formally,

f is a function
if and only if
f is a relation, and
$(x, y) \in f$ and $(x, z) \in f$ implies $y = z$.

By the domain of a relation we mean the set of all first coordinates of ordered pairs of the relation. If f is a relation we abbreviate domain f by dmn f, and define formally,

$$\text{dmn } f = \{x \mid \text{for some } y,\ (x, y) \in f\}.$$

By the range of a relation we mean the set of all second coordinates of ordered pairs of the relation. If f is a relation we abbreviate range f by rng f, and define formally,

$$\text{rng } f = \{y \mid \text{for some } x, (x, y) \in f\}.$$

Observe that if f is a function and if $x \in \text{dmn } f$, then $f(x)$ is that unique element such that $(x, f(x)) \in f$. If f is a function and if $x \notin \text{dmn } f$, then $f(x)$ is meaningless. Finally if f is not a function then $f(x)$ is meaningless for every x.

1.5 Superpositions and Inverses

If f and g are functions we shall frequently superimpose f upon g to manufacture that function which yields for each x in its domain $f(g(x))$. Clearly the domain of the superposition of f upon g consists of those $x \in \text{dmn } g$ such that $g(x) \in \text{dmn } f$. We shall denote the superposition of f upon g by $f \circ g$. Observe that $f \circ g$ consists of all ordered pairs (x, y) such that for some z, $(x, z) \in g$ and $(z, y) \in f$. With this in mind we can consider $f \circ g$ when f and g are merely relations, and define formally,

$$f \circ g = \{(x, y) \mid \text{for some } z, (x, z) \in g \text{ and } (z, y) \in f\}.$$

If f is a relation, then by the inverse of f we mean the relation obtained by reversing the coordinates of all ordered pairs of f. We denote the inverse of f by inv f and define formally,

$$\text{inv } f = \{(y, x) \mid (x, y) \in f\}.$$

If f is a function such that inv f is also a function, then f is usually called one-to-one or univalent. We shall prefer the term univalent, and define formally,

$$f \text{ is univalent}$$
$$\text{if and only if}$$
$$f \text{ is a function and inv } f \text{ is a function.}$$

1.6 Restrictions

If f is a relation and B is any set we frequently want to consider only those ordered pairs of f whose first coordinates are elements of B. This set is called the restriction of f to B, and we denote it by $f|B$. Formally,

$$f|B = f \cap \{(x, y) \mid x \in B\}.$$

1.7 Cartesian Products

For any sets A and B, the Cartesian product of A and B consists of all ordered pairs (a, b) such that $a \in A$ and $b \in B$. We denote the Cartesian product of A and B by $(A \times B)$, and define formally,

$$(A \times B) = \{(a, b) \mid a \in A \text{ and } b \in B\}.$$

Closely related to the Cartesian product $(A \times B)$ is the set of all sets of exactly two elements, at least one of which is an element of A and at least one of which is an element of B. We denote this set by $(A : B)$ and observe that each element of $(A : B)$ is a set of the type $\{a, b\}$ such that $a \in A$ and $b \in B$ and $a \neq b$. Thus, formally,

$$(A : B) = \{\{a, b\} \mid a \in A \text{ and } b \in B \text{ and } a \neq b\}.$$

1.8 Some Special Symbols

In this section we introduce some special symbols to be used throughout the book

(1) $R = \{x \mid x \text{ is a real number}\}$.
(2) $\omega = \{x \mid x \text{ is a positive integer}\}$.
(3) For each $n \in \omega$

$$[n] = \{x \mid x \in \omega \text{ and } x \leq n\}.$$

Thus, for each $n \in \omega$, $[n]$ is the set consisting of the first n positive integers, $1, 2, \ldots, n$.

(4) If A is a finite set, then pA is the number of elements of A.
(5) For any set A,
$$I_A = \{(x, x) \mid x \in A\}.$$

Thus, if A is non-empty, then I_A is the identity function with domain A such that $I_A(x) = x$ for each $x \in A$.

1.9 Finite Sequences

By a finite sequence we mean a function whose domain is $[n]$ for some $n \in \omega$. If f is a finite sequence and if $j \in \operatorname{dmn} f$, we adhere to the usual custom and frequently write f_j instead of $f(j)$. Note that f and $\operatorname{dmn} f$ each have exactly pf elements, and thus, f_{pf} is a convenient way to describe the last element of the range of f.

1.10 Networks

Eventually, we shall define a network in its most abstract form. To motivate our abstract formulation, recall that the usual definition of a network requires that it should be a collection of straight line segments and points in three-dimensional space. The line segments are called branches and the points are called vertices. Two branches of the network can intersect only at a common end point of each branch. The two end points of each branch must be included in the set of vertices of the network, but it is not necessary that every line segment joining two vertices of the network be included in the set of branches of the network. In topological terms such a network could be described as a one-dimensional complex in three-dimensional space.

Observe that if the vertices of a network are known, then the network can be completely specified by exhibiting those pairs of vertices which are end points of a branch of the network. Thus, the network is specified by a set of vertices V, and certain sets of the type $\{a, b\}$ such that $a \in V$ and $b \in V$ and $a \neq b$. Each such set $\{a, b\}$ with $a \in V$ and $b \in V$ and $a \neq b$ is an element of $(V : V)$. This leads to a description of a network as an ordered pair of sets (V, S) such that $S \subset (V : V)$.

With this abstract formulation of a network (V, S) there is no need to require that each vertex of V is a point in three-dimensional space. Any set V will suffice for the set of vertices, although we shall impose the restriction that V must be finite and non-empty. The set S, however, may be empty. For convenience, we shall still refer to the elements of S as branches.

Thus, we have our abstract definition of a network.

(V, S) is a network

if and only if

(1) V is finite and $V \neq 0$.

(2) $S \subset (V : V)$.

Consider two networks, (V, S) and (V', S'). By taking unions and intersections we manufacture the ordered pairs $(V \cup V', S \cup S')$ and $(V \cap V', S \cap S')$, and ask whether each of these is a network. It is always true that $(V \cup V', S \cup S')$ is a network, but $(V \cap V', S \cap S')$ is a network if and only if $V \cap V'$ is non-empty. We cite these facts in the following Theorem 1, but omit the trivial proof.

Theorem 1. *Let (V, S) be a network and let (V', S') be a network. Then,*

(1) *$(V \cup V', S \cup S')$ is a network.*

(2) *$(V \cap V', S \cap S')$ is a network if and only if $V \cap V' \neq 0$.*

Suppose that (V, S) is a network and L is a non-empty set of branches of S. Note that σL consists of all vertices of the branches of L. Thus, it is clear that $(\sigma L, L)$ is a network. This fact is stated in the following Theorem 2, but the trivial proof is omitted.

Theorem 2. *Let (V, S) be a network. Let $L \subset S$ such that $L \neq 0$. Then,*

$(\sigma L, L)$ is a network.

On the other hand, suppose that S is a set such that $(\sigma S, S)$ is a network. Then, $S \neq 0$, for otherwise, $\sigma S = \sigma 0 = 0$, contradicting the fact that $(\sigma S, S)$ is a network. We cite this fact as the following Theorem 3, but omit the proof.

Theorem 3. *Let $(\sigma S, S)$ be a network. Then,*

$S \neq 0$.

1.11 Geometrical Realization of a Network

Let (V, S) be a network. By selecting an appropriate set of pV points in three-dimensional space in one-to-one correspondence with the elements of V, and by drawing those straight line segments connecting those pairs of points in three-dimensional space corresponding to the pairs of vertices of S, we can manufacture a geometrical realization of the abstract network (V, S). Thus, a vertex of V is an element of a pair of vertices of S if and only if the corresponding point of the geometrical realization is an end point of the corresponding branch of the geometrical realization.

Observe, however, that the pV points in three-dimensional space must be selected with a little care. For example, if all pV points are confined to one plane, then, except in certain special cases, no geometrical realization of (V, S) can be constructed using these pV points as vertices. However, the problem of the proper selection of the pV points is topological in nature, and is not vital to the circuit theory to be developed in this book.

For our purposes in this book it suffices to know that a geometrical realization of any network can always be constructed. We shall frequently use a geometrical realization of a network to give the reader some geometrical insight into the abstract concepts under consideration.

1.12 Subnetworks

If (V, S) is a network, then a subnetwork of (V, S) is a network (M, N) such that each vertex of M is a vertex of V and each branch of N is a branch is S. Formally,

(M, N) is a subnetwork of (V, S)

if and only if

(1) (V, S) is a network.

(2) (M, N) is a network.

(3) $M \subset V$ and $N \subset S$.

We shall write

$$(M, N) \in (V, S)$$

as an abbreviation for the statement that (M, N) is a subnetwork of (V, S).

Note that the relation of being a subnetwork is transitive. Also, if (V, S) is a network and if M is any subset of S, then (V, M) is a subnetwork of (V, S). We cite these two facts formally as a Theorem, but omit the trivial proof.

Theorem.

(1) $(K, L) \in (M, N) \in (V, S)$ *implies* $(K, L) \in (V, S)$.

(2) *If* (V, S) *is a network and* $M \subset S$, *then* $(V, M) \in (V, S)$.

1.13 Degree of a Vertex

Let (V, S) be a network and let $x \in V$. Consider any geometrical realization of (V, S) and let x' be that point of the geometrical realization corresponding to x. We ask how many branches of the geometrical realization have x' as an endpoint. In terms of the abstract network (V, S), this number is equal to the number of elements of S of which x is an element.

We generalize this by considering any subset M of S, and finding the number of elements of M of which x is an element. This number is called the degree of x with respect to M, and we denote it by $\deg_M(x)$. Formally,

$$\deg_M(x) = p\{y \mid x \in y \in M\}.$$

1.14 Path in a Network

Let (V, S) be a network and let $x \in V$ and let $y \in V$. Consider any geometrical realization of (V, S) and let x' and y' be the points of the geometrical realization corresponding to x and y respectively. By a path from x' to y' in the geometrical realization, we mean a sequence of branches of the geometrical realization such that x' is an end point of the first branch of the sequence, y' is an endpoint of the last branch of the sequence, and, with the exception of the first branch of the sequence, each branch has an endpoint in common with the branch immediately preceding it.

In terms of the abstract network (V, S), a path from x to y becomes a finite sequence f with $\operatorname{rng} f \subset S$ such that $x \in f_1$, $y \in f_{pf}$, and $f_i \cap f_{i-1} \neq 0$ for each $i \in \operatorname{dmn} f$ with $1 < i$. Observe, however, that V, the set of vertices of the network (V, S), is of no significance in the description of the path f. In fact, we can consider a path in any set S, provided that S is a set of branches. In the formal definition which follows we guarantee that the set S is a set of branches by demanding that $(\sigma S, S)$ is a network.

f is a path from x to y in S

if and only if

(1) $(\sigma S, S)$ is a network.

(2) $x \in \sigma S$ and $y \in \sigma S$.

(3) f is a finite sequence.

(4) $\operatorname{rng} f \subset S$.

(5) $x \in f_1$ and $y \in f_{pf}$.

(6) $1 < i \in \operatorname{dmn} f$ implies $f_i \cap f_{i-1} \neq 0$.

1.15 Proper Path in a Network

Let f be a path from x to y in S with $x \neq y$. Consider any geometrical realization of $(\sigma S, S)$ and let x' and y' be the points of the geometrical realization corresponding to x and y respectively. In the geometrical realization examine the path from x' to y' corresponding to f. Now, this path in the geometrical realization may have some objectionable features. In particular, some branches may be repeated, some branches may be completely extraneous, and the path may cross itself.

Returning to the abstract network $(\sigma S, S)$, we want to define a proper path from x to y in such a way that these objectionable features will be eliminated from the path in the geometrical realization. By requiring that f is univalent we can guarantee that no branch is repeated in the path in the geometrical realization. Extraneous branches and self-crossings can be eliminated from the path in the geometrical realization if we demand that $\deg_{\operatorname{rng} f}(x) = 1$, $\deg_{\operatorname{rng} f}(y) = 1$, and $\deg_{\operatorname{rng} f}(z) = 2$ for each $z \in \sigma \operatorname{rng} f$ with $z \neq x$ and $z \neq y$. Formally,

f is a proper path from x to y in S

if and only if

(1) f is a path from x to y in S.

(2) $x \neq y$.

(3) f is univalent.

(4) $\deg_{\operatorname{rng} f}(x) = 1$ and $\deg_{\operatorname{rng} f}(y) = 1$.

(5) $z \in \sigma \operatorname{rng} f$ and $z \neq x$ and $z \neq y$ implies
$\deg_{\operatorname{rng} f}(z) = 2$.

Suppose that f is a proper path from x to y in S. By Theorem 2 of 1.10, $(\sigma \operatorname{rng} f, \operatorname{rng} f)$ is a network, and later we shall find it useful to compare the number of vertices of this network $(\sigma \operatorname{rng} f, \operatorname{rng} f)$ with the number of branches of the same network.

To do this, suppose that $\operatorname{rng} f$ has exactly k elements. By using the definition of a proper path it is easy to manufacture a univalent finite sequence u with domain $[k + 1]$ such that $\operatorname{rng} u = \sigma \operatorname{rng} f$, $u_1 = x$, $u_{k+1} = y$, and $\{u_i, u_{i+1}\} = f_i$ for each $i \in [k]$. Since u is univalent and $\operatorname{rng} u = \sigma \operatorname{rng} f$, it is clear that $\sigma \operatorname{rng} f$ has exactly $k + 1$ elements. Thus, the network $(\sigma \operatorname{rng} f, \operatorname{rng} f)$ has exactly one more vertex than branches.

These facts, including the existence of the finite sequence u, are cited in the following Theorem, but we omit the proof, since it depends only trivially upon the definition of a proper path.

Theorem. *Let f be a proper path from x to y in S. Suppose that $p \operatorname{rng} f = k$.*
Then,

   (1) *For some u,*
       (i)   *u is a univalent finite sequence.*
       (ii)   *dmn $u = [k + 1]$ and $\operatorname{rng} u = \sigma \operatorname{rng} f$.*
       (iii)   *$u_1 = x$ and $u_{k+1} = y$.*
       (iv)   *$i \in [k]$ implies $\{u_i, u_{i+1}\} = f_i$.*
   (2) *$p \sigma \operatorname{rng} f = k + 1 = 1 + p \operatorname{rng} f$.*

1.16 Reduction of a Path to a Proper Path

Every path in a network can be reduced to a proper path. More precisely, let f be a path from x to y in S with $x \neq y$. Thus, there exists g such that g is a proper path from x to y in S and $\operatorname{rng} g \subset \operatorname{rng} f$. To manufacture g, let m be the smallest positive integer such that for some h, h is a path from x to y in S, $\operatorname{rng} h \subset \operatorname{rng} f$, and dmn $h = [m]$; let g be such a path corresponding to this minimal m.

This fact is stated formally as a Theorem.

Theorem. *Let f be a path from x to y in S such that $x \neq y$. Then,*

for some g,

(1) *g is a proper path from x to y in S.*

(2) *rng $g \subset$ rng f.*

Proof: Let

$$m = \min \{k \mid \text{for some } h, \ h \text{ is a path from } x \text{ to } y \text{ in } S \text{ and}$$
$$\text{rng } h \subset \text{rng } f \text{ and dmn } h = [k]\}.$$

Produce g such that g is a path from x to y in S, rng $g \subset$ rng f, and dmn $g = [m]$. By the minimality of m it is easy to show that g is a proper path from x to y in S.

1.17 Connected Networks

We want to define a connected network in such a way that every geometrical realization shall be a connected subset, in the topological sense, of three-dimensional space. Recall that a topological space is connected if and only if it cannot be written as the union of two non-empty, disjoint open subsets of the space. Analogously, we shall say that an abstract network is connected if and only if it cannot be written as the union of two disjoint subnetworks. Formally,

> (V, S) is a connected network
>
> if and only if
>
> (V, S) is a network, and,
>
> there do not exist (V', S') and (V'', S'') such that
>
> (1) $(V', S') \in (V, S)$ and $(V'', S'') \in (V, S)$.
>
> (2) $V = V' \cup V''$ and $S = S' \cup S''$.
>
> (3) $V' \cap V'' = 0$.

1.18 Isolated Vertices

By an isolated vertex of a network (V, S) we mean a vertex $x \in V$ such that x is not an element of any branch of S. Thus, a vertex $x \in V$ is isolated if and only if $x \notin \sigma S$. Equivalently, in terms of degree, a vertex $x \in V$ is isolated if and only if $\deg_S(x) = 0$. Formally,

x is an isolated vertex of (V, S)

if and only if

(1) (V, S) is a network.

(2) $x \in V$.

(3) $\deg_S(x) = 0$.

Observe that for any x, $(\{x\}, 0)$ is a connected network and x is an isolated vertex of $(\{x\}, 0)$. However, in the following two Theorems we point out that this is the only situation which can produce an isolated vertex of a connected network.

Theorem 1 discusses the case of a connected network (V, S) with $S = 0$, and points out that $(V, 0)$ is a connected network if and only if V consists of exactly one element. We omit the trivial proof.

Theorem 2 discusses the case of a connected network (V, S) with $S \neq 0$, and points out that such a network has no isolated vertices.

Theorem 1. *$(V, 0)$ is a connected network if and only if $pV = 1$.*

Theorem 2. *Let (V, S) be a connected network such that $S \neq 0$. Let $x \in V$.*
Then,
    x is not an isolated vertex of (V, S).

Proof: Suppose that x is an isolated vertex of (V, S). Then, $S \subset (V - \{x\} : V - \{x\})$, and,

$$(V - \{x\}, S) \in (V, S) \text{ and } (\{x\}, 0) \in (V, S),$$
$$V = (V - \{x\}) \cup \{x\} \text{ and } S = S \cup 0,$$
$$(V - \{x\}) \cap \{x\} = 0,$$

contradicting the fact that (V, S) is a connected network. The proof is complete.

If (V, S) is a network, then we always have trivially $\sigma S \subset V$. If, in addition, there are no isolated vertices of (V, S), then it is also true that $V \subset \sigma S$. Thus, in a network (V, S) with no isolated vertices, $V = \sigma S$. We cite this fact in the following Theorem 3. Finally, by combining Theorem 3 and Theorem 2, we point out in Theorem 4 that if (V, S) is a connected network with $S \neq 0$, then $V = \sigma S$.

Theorem 3. *Let (V, S) be a network. Suppose that*
    $\{x \mid x$ is an isolated vertex of $(V, S)\} = 0$.
Then,
    $V = \sigma S$.

Proof: Trivially $\sigma S \subset V$. To show $V \subset \sigma S$, let $x \in V$. Since x is not an isolated vertex of (V, S), produce y such that $x \in \{x, y\} \in S$, implying that $x \in \sigma S$.

Theorem 4. *Let (B, S) be a connected network such that $S \neq 0$. Then,*

$$V = \sigma S.$$

Proof: By Theorem 2, $\{x \mid x$ is an isolated vertex of $(V, S)\} = 0$. By Theorem 3, $V = \sigma S$.

1.19 Connected Sets of Branches

Ultimately, we shall apply our abstract network to circuit theory. Now, circuits with isolated nodes are of no practical, and little academic interest. Thus, we shall almost exclusively concern ourselves with networks (V, S) with no isolated vertices. But in Theorem 3 of 1.18 we point out that in such a network (V, S) with no isolated vertices, $V = \sigma S$. Hence, in such a network, a knowledge of S specifies V completely, and we can dispense with the secretarial problem of using the extra letter V to describe the network.

In 1.17 we defined a connected network. The preceding paragraph suggests that a simplification in the notation might be achieved by defining the concept of a connected set of branches S. This is accomplished by stating that a set S is connected if and only if $(\sigma S, S)$ is a connected network. The requirement that $(\sigma S, S)$ is a network guarantees that S is a set of branches. Formally,

S is connected

if and only if

$(\sigma S, S)$ is a connected network.

The tactics exhibited above will be repeated frequently throughout the book. We shall introduce concepts with reference to a network structure, but then find that it is much simpler to consider these concepts with reference to a branch structure.

It is convenient to characterize the connectedness of a set of branches S by the non-existence of certain subsets $S' \subset S$ and $S'' \subset S$ with properties analogous to the network properties listed in 1.17. This is accomplished in the following Theorem.

Theorem. *S is connected if and only if $(\sigma S, S)$ is a network, and there do not exist S' and S'' such that*

(1) $S' \subset S$ *and* $S'' \subset S$ *and* $S' \neq 0$ *and* $S'' \neq 0$.

(2) $S = S' \cup S''$.

(3) $\sigma S' \cap \sigma S'' = 0$.

Proof: Use 1.17 and the fact that $\sigma(S' \cup S'') = \sigma S' \cup \sigma S''$ for any sets S' and S''.

1.20 Path Connected Set of Branches

Recall that an arcwise connected topological space is a space such that each two points of the space can be joined by an arc in the space. Consider now a network $(\sigma S, S)$ and a geometrical realization of it. Because σS is finite, the geometrical realization is an arcwise connected subset. Thus, referring back to the abstract network $(\sigma S, S)$ we might expect S to be a connected set of branches if and only if some condition, analogous to arcwise connectedness in the geometrical realization, is valid in the abstract network.

We call such a condition in the abstract network, path connectedness. By a path connected set S we mean a set S such that $(\sigma S, S)$ is a network and such that between each two vertices of the network $(\sigma S, S)$ a path in S exists. Formally,

> S is path connected
>
> if and only if
>
> $(\sigma S, S)$ is a network, and
>
> $x \in \sigma S$ and $y \in \sigma S$ and $x \neq y$ implies
>
> for some f,
>
> f is a path from x to y in S.

Note that by the Theorem of 1.16, we can replace "f is a path from x to y in S" in the above definition by "f is a proper path from x to y in S".

As anticipated, the following Theorem points out that a set of branches is connected if and only if it is path connected.

Theorem. *S is connected if and only if S is path connected.*
Proof: Let S be connected. Let $x \in \sigma S$ and let $y \in \sigma S$ with $x \neq y$. Let

$$Q = \{t \mid \text{for some } f, f \text{ is a path from } x \text{ to } t \text{ in } S\}.$$

It suffices to show that $y \in Q$.

Suppose that $y \notin Q$. We shall manufacture sets $S' \subset S$ and $S'' \subset S$ contradicting the connectedness of S as characterized by the Theorem of 1.19. This will suffice to complete the proof in this direction. First let

$$G = \{f \mid \text{for some } t, f \text{ is a path from } x \text{ to } t \text{ in } S\}.$$

Then, let

$$S' = \operatorname{rng} \sigma G.$$
$$S'' = S - S'.$$

It is immediate that $S' \subset S$ and $S'' \subset S$. We now assert that $S' \neq 0$ and $S'' \neq 0$.

To show that $S' \neq 0$, produce v such that $\{x, v\} \in S$ and note that $\{(1, \{x, v\})\}$ is a path from x to v in S, implying that $\{(1, \{x, v\})\} \in G$ and $\{x, v\} \in \operatorname{rng} \sigma G = S'$.

To show that $S'' \neq 0$, produce u such that $\{u, y\} \in S$. We shall show that $\{u, y\} \in S''$. Suppose that $\{u, y\} \notin S''$, implying that $\{u, y\} \in S'$. Produce g and j such that $(j, \{u, y\}) \in g \in G$. Then, $j \in \operatorname{dmn} g$, $\{u, y\} = g_j$, and $g|[j]$ is a path from x to y in S, implying that $y \in Q$ which is false.

Thus, $S' \neq 0$ and $S'' \neq 0$.

It is immediate that $S = S' \cup S''$.

Referring to the Theorem of 1.19 we see that to contradict the fact that S is connected we need only show that $\sigma S' \cap \sigma S'' = 0$.

Suppose that $z \in \sigma S' \cap \sigma S''$.

Since $z \in \sigma S'$, produce b such that $z \in \{b, z\} \in S'$.

Then, produce k and h such that $(k, \{b, z\}) \in h \in G$.

Note that $k \in \operatorname{dmn} h$, $\{b, z\} = h_k$, and $h|[k]$ is a path from x to z in S.

But since $z \in \sigma S''$, produce c such that $\{z, c\} \in S''$.

Observe that $h|[k] \cup \{(k + 1, \{z, c\})\}$ is a path from x to c in S, implying that

$$(k + 1, \{z, c\}) \in h|[k] \cup \{(k + 1, \{z, c\})\} \in G,$$

and $\{z, c\} \in \operatorname{rng} \sigma G = S'$. This contradicts the fact that $\{z, c\} \in S''$.

Thus, we must have $\sigma S' \cap \sigma S'' = 0$.

By the Theorem of 1.19, S is not connected, which is false, and the proof in this direction is complete.

In the other direction, suppose that S is path connected but not connected.

Produce sets T' and T'' such that

$$T' \subset S \text{ and } T'' \subset S \text{ and } T' \neq 0 \text{ and } T'' \neq 0,$$
$$S = T' \cup T'',$$
$$\sigma T' \cap \sigma T'' = 0.$$

Produce $\{r, w\} \in T'$ and $\{d, s\} \in T''$.

Then, $r \in \sigma T'$ and $s \in \sigma T''$, implying that $r \neq s$.

Produce f such that f is a path from r to s in S.

Let

$$m = \max \{i | i \in \operatorname{dmn} f \text{ and } f_i \in T'\}.$$

Then, it is easy to check that

$$f_m \in T',$$
$$f_{m+1} \in T'',$$
$$0 \neq f_m \cap f_{m+1} \subset \sigma T' \cap \sigma T'',$$

contradicting the fact that $\sigma T' \cap \sigma T'' = 0$. The proof is complete.

1.21 Union of Connected Sets of Branches

If two connected sets of branches have one or more vertices in common, their union is a connected set of branches. We cite this fact as the following Theorem.

Theorem. *Let S be connected. Let S' be connected. Suppose that $\sigma S \cap \sigma S' \neq 0$.*
Then,
$S \cup S'$ is connected.

Proof: It suffices to show that $S \cup S'$ is path connected. Thus, we must first show that $(\sigma\,(S \cup S'),\, S \cup S')$ is a network.

We know that $(\sigma S,\, S)$ is a network and $(\sigma S',\, S'')$ is a network. By (1) of Theorem 1 of 1.10, $(\sigma S \cup \sigma S',\, S \cup S')$ is a network. Since $\sigma\,(S \cup S') = \sigma S \cup \sigma S'$, $(\sigma\,(S \cup S'),\, S \cup S')$ is a network.

Now let $x \in \sigma\,(S \cup S') = \sigma S \cup \sigma S'$ and let $y \in \sigma\,(S \cup S') = \sigma S \cup \sigma S'$ such that $x \neq y$. We may assume that $x \in \sigma S$. Now, if $y \in \sigma S$, then there exists a path from x to y in S which is also a path from x to y in $S \cup S'$, and we are through. Thus, we may assume that $y \in \sigma S' - \sigma S$. Similarly, our assumption that $y \in \sigma S'$ lets us specialize x such that $x \in \sigma S - \sigma S'$.

Let $z \in \sigma S \cap \sigma S'$, and note that $z \neq x$ and $z \neq y$. By producing a path from x to z in S and also producing a path from z to y in S' it is easy to manufacture from these two paths a third path from x to y in $S \cup S'$.

1.22 Connectedness of Paths

Consider a path f joining two vertices of a set of branches. Then the set of branches rng f is path connected and connected. To show that rng f is path connected, let $u \in \sigma\,\text{rng}\,f$ and let $v \in \sigma\,\text{rng}\,f$ with $u \neq v$. It is easy to manufacture a subsequence g of f such that g is a path from u to v in rng f. In the following Theorem we cite the connectedness of rng f, but omit the proof.

Theorem. *Let f be a path from x to y in S.*
Then,
rng f is connected.

1.23 Component of a Set of Branches

Recall that a component of a topological space is a maximal connected subset of the space. Analogously, we define a component of a set of branches S as a maximal connected subset of S. Formally,

A is a component of S

if and only if

(1) $(\sigma S, S)$ is a network.

(2) $A \subset S$.

(2) A is connected.

(4) $A \subset B \subset S$ and B is connected
      implies
      $A = B$.

Consider a network $(\sigma S, S)$ and any geometrical realization of it. The preceding definition guarantees that A is a component of S if and only if the subset corresponding to A in the geometrical realization is a component, in the topological sense, of the geometrical realization.

If A is a component of S and if B is a connected subset of S with a vertex in common with A, then the maximality of A guarantees that $B \subset A$. We cite this fact as the following Theorem. As a Corollary we obtain the fact that two components of a set of branches are either equal or have no vertices in common.

Theorem. *Let A be a component of S. Let $B \subset S$ such that B is connected. Suppose that $\sigma A \cap \sigma B \neq 0$.*
Then,
   $B \subset A$.

Proof: By the Theorem of 1.21, $A \cup B$ is connected. But $A \subset A \cup B \subset S$. By the maximality of A, $A \cup B \subset A$, implying that $B \subset A$.

Corollary. *Let A be a component of S. Let B be a component of S. Then $A = B$ or $\sigma A \cap \sigma B = 0$.*

Proof: It suffices to assume that $\sigma A \cap \sigma B \neq 0$, and show $A = B$. But by using the preceding Theorem twice, $B \subset A$ and $A \subset B$, implying that $A = B$.

1.24 Existence of Components

If B is a connected subset of a set of branches S, then there exists exactly one component A of S such that $B \subset A$. This unique component A of S is easily manufactured as the union of all those connected subsets of S which include B as a subset. The existence and uniqueness of this component A of S is established in the following Theorem.

Theorem. *Let $(\sigma S, S)$ be a network. Let $B \subset S$ such that B is connected. Then,*
   $p\{F \,|\, B \subset F$ and F is a component of $S\} = 1$.

Proof: Let

$$M = \{F \mid B \subset F \text{ and } F \text{ is a component of } S\}$$
$$K = \{F \mid B \subset F \subset S \text{ and } F \text{ is connected}\}$$
$$A = \sigma K.$$

To complete the proof we shall show that $M = \{A\}$.

Note first that $B \in K$ implying that $B \subset \sigma K = A$.

Also, since K is finite, a trivial inductive extension of the Theorem of 1.21 shows that A is connected.

Thus, $A \in M$, implying that $\{A\} \subset M$.

To show that $M \subset \{A\}$, suppose that $F \in M$, implying that $B \subset F$ and F is a component of S. Also, we have shown $A \in M$, implying that $B \subset A$ and A is a component of S. Thus, $B \subset F \cap A$, implying that

$$0 \neq \sigma B \subset \sigma(F \cap A) \subset \sigma F \cap \sigma A.$$

By the Corollary of 1.23, $F = A$, completing the proof.

1.25 Partition into Components

Let (V, S) be a network. S is partitioned into disjoint subsets by the set of all components of S and S is equal to the union of the set of all components of S. Furthermore, all non-isolated vertices of V can be realized by taking a second union of the set of all components of S. However, the isolated vertices of V cannot be extracted from the set of all components of S. These facts are pointed out in the following Theorem.

Theorem. *Let (V, S) be a network.*
Then,

(1) $S = \sigma\{A \mid A \text{ is a component of } S\}$.

(2) $V = \sigma\sigma\{A \mid A \text{ is a component of } S\} \cup \{x \mid x \in V \text{ and } \deg_S(x) = 0\}$.

Proof: Let

$$M = \{A \mid A \text{ is a component of } S\}.$$

To prove (1) it is immediate that $\sigma M \subset S$. To show that $S \subset \sigma M$, let $\{x, y\} \in S$ and observe that $\{\{x, y\}\} \subset S$ and $\{\{x, y\}\}$ is connected. By the Theorem of 1.24, produce $B \in M$ such that $\{\{x, y\}\} \subset B$. Thus, $\{x, y\} \in B \in M$, implying that $\{x, y\} \in \sigma M$.

To prove (2) it is immediate that
$\sigma\sigma M \cup \{x \mid x \in V \text{ and } \deg_S(x) = 0\} \subset V$.

To show that $V \subset \sigma\sigma M \cup \{x \mid x \in V \text{ and } \deg_S(x) = 0\}$, suppose that $v \in V$ and $\deg_S(v) \neq 0$.

Produce u such that $v \in \{v, u\} \in S$, implying that $v \in \sigma S$. By (1), $v \in \sigma\sigma M$.

1.26 Removal of a Branch

Consider a connected set of branches S and a branch $\{x, y\} \in S$. We remove the branch $\{x, y\}$ from S and examine the remaining set of branches $S - \{\{x, y\}\}$. In particular, we assess the number of components of $S - \{\{x, y\}\}$.

In the following Theorem 1 we point out that the number of components of $S - \{\{x, y\}\}$ cannot exceed 2. Next, in Theorem 2 we show that in the special case where either of the vertices x or y meets only one branch of S, the number of components of $S - \{\{x, y\}\}$ cannot exceed 1.

These two Theorems can be proved more efficiently by preceding them with a Lemma.

Lemma. *Let S be connected. Let $\{x, y\} \in S$. Let M be a component of $S - \{\{x, y\}\}$ such that $x \in \sigma M$. Let N be a component of $S - \{\{x, y\}\}$ such that $M \neq N$.*
Then,

$$y \in \sigma N.$$

Proof: Produce $z \in \sigma N$ such that $z \neq y$ and note that $z \neq x$, because $x \in \sigma M$ and $\sigma M \cap \sigma N = 0$. Since S is connected, produce f such that f is a proper path from z to x in S.

Now, if $\{x, y\} \notin \operatorname{rng} f$, then $\operatorname{rng} f \subset S - \{\{x, y\}\}$, $\operatorname{rng} f$ is connected, and $z \in \sigma \operatorname{rng} f \cap \sigma N$. By the Theorem of 1.23, $x \in \sigma \operatorname{rng} f \subset \sigma N$, which is false since $x \in \sigma M$ and $\sigma M \cap \sigma N = 0$. Thus, we must have $\{x, y\} \in \operatorname{rng} f$.

Furthermore, since f is a proper path, $\{x, y\} = f_{pf}$. But then, $f \mid [pf - 1]$ is a proper path from z to y in $S - \{\{x, y\}\}$, $\operatorname{rng} (f \mid [pf - 1])$ is connected, and $z \in \sigma \operatorname{rng} (f \mid [pf - 1]) \cap \sigma N$. By the Theorem of 1.23, $y \in \sigma \operatorname{rng} (f \mid [pf - 1]) \subset \sigma N$, completing the proof.

Theorem 1. *Let S be connected. Let $\{x, y\} \in S$.*
Then,

$$p \{A \mid A \text{ is a component of } S - \{\{x, y\}\}\} \leq 2.$$

Proof: Suppose that $p\{A \mid A$ is a component of $S - \{\{x, y\}\}\} \geq 3$. Now, if $\deg_S(x) = 1$ and $\deg_S(y) = 1$, observe that $\{\{x, y\}\}$ is a component of S but $\{\{x, y\}\} \neq S$, contradicting the fact S is connected. Thus, $\deg_S(x) > 1$ or $\deg_S(y) > 1$.
We assume that $\deg_S(x) > 1$.

Since $\deg_S(x) > 1$, produce M such that M is a component of $S - \{\{x, y\}\}$ and $x \in \sigma M$. Also produce N and N' such that N is a component of $S - \{\{x, y\}\}$, N' is a component of $S - \{\{x, y\}\}$, $M \neq N$, $M \neq N'$, and $N \neq N'$. Apply the preceding Lemma twice to deduce that $y \in \sigma N \cap \sigma N'$, which is false.

Theorem 2. *Let S be connected. Let $\{x, y\} \in S$. Suppose that* $\deg_S(y) = 1$. *Then,*

$$p\left\{A \mid A \text{ is a component of } S - \{\{x, y\}\}\right\} \leq 1.$$

Proof: Suppose that $p\{A \mid A$ is a component of $S - \{\{x, y\}\}\} \geq 2$. Now, if $\deg_S(x) = 1$, then $\{\{x, y\}\}$ is a component of S, but $\{\{x, y\}\} \neq S$, which is false. Thus, $\deg_S(x) > 1$.

Produce M such that M is a component of $S - \{\{x, y\}\}$ and $x \in \sigma M$. Produce N such that N is a component of $S - \{\{x, y\}\}$ and $M \neq N$. By the preceding Lemma, $y \in \sigma N$. Produce z such that $y \in \{y, z\} \in N$. Thus, $z \in \sigma N$ and $z \neq x$, because $x \in \sigma M$ and $\sigma M \cap \sigma N = 0$. But then $\deg_S(y) \geq 2$, which is false.

CHAPTER TWO

Loops, Trees, and Cut Sets

2.0 Introduction

Now that we are experts on the general subject of connected networks and connected sets of branches, we focus our attention upon two particular types of connected sets of branches, loops and trees. In addition, in this Chapter we consider cut sets, which are sets of branches very closely associated with the concept of connectedness.

2.1 Loop in a Network

Consider a network (V, S) and a geometrical realization of it. We want to define a loop L in (V, S) as a set of branches of S such that the branches corresponding to L in the realization shall form what is usually considered to be a loop in three-dimensional space. By a loop in three-dimensional space a mathematician means any subset which is homeomorphic to the circumference of a circle.

There appears to be a choice available for our definition of a loop L in the abstract network (V, S). For example, as our definition we could require the connectedness of the set of branches of L, plus the existence, for each vertex $x \in \sigma L$ of exactly one univalent path in L, beginning and ending at x. Alternatively, as our definition we could require the connectedness of L, plus the existence for each branch $\{x, y\} \in L$ of exactly one proper path from x to y in $L - \{\{x, y\}\}$.

These suggested definitions of a loop and other equivalent formulations of the definition using the existence of paths appear, to this writer, rather clumsy and inelegant. A simpler requirement for the definition of a loop, in addition to the connectedness of the set of branches L, is the fact that each vertex of L must meet exactly two branches of L. In terms of degree, this latter condition reduces to the statement that $\deg_L(x) = 2$ for each $x \in \sigma L$. We shall adopt this simple condition involving degree together with the connectedness of L as our definition for L to be a loop in (V, S). Formally,

L is a loop in (V, S)

if and only if

(1) (V, S) is a network.

(2) $L \subset S$.

(3) L is connected.

(4) $x \in \sigma L$ implies $\deg_L(x) = 2$.

2.2 Loops

If L is a loop in a network (V, S), a discerning reader will see that the network (V, S) in which the loop L is imbedded is of no great significance. Thus, it is appropriate to provide an intrinsic definition of a loop, which is independent of the network in which the loop is imbedded. This is accomplished by stating that any set L is a loop if and only if L is a loop in the network $(\sigma L, L)$. Formally,

L is a loop

if and only if

L is a loop in $(\sigma L, L)$.

2.3 Subloops of a Loop

Let L be a loop. Examination of a geometrical realization of $(\sigma L, L)$ discloses that there exist no proper subloops of L. More precisely, there exists no M such that $M \subset L$, M is a loop, and $M \neq L$. We exhibit this fact, which will be used frequently, as the following Theorem.

Theorem. *Let L be a loop. Let M be a loop such that $M \subset L$. Then,*
$M = L$.

Proof: Suppose that $M \neq L$, implying that $L - M \neq 0$. Now, if $\sigma M \cap \sigma(L - M) = 0$, observe that

$$M \subset L \text{ and } L - M \subset L \text{ and } M \neq 0 \text{ and } L - M \neq 0,$$
$$L = M \cup (L - M),$$
$$\sigma M \cap \sigma(L - M) = 0,$$

contradicting the fact that L is connected.
Thus, $\sigma M \cap \sigma(L - M) \neq 0$.

Produce $x \in \sigma M \cap \sigma(L - M)$. Since M is a loop and $\deg_M(x) = 2$, produce u and v such that $x \in u \in M \subset L$ and $x \in v \in M \subset L$ and $u \neq v$. Since $x \in \sigma(L - M)$, produce w such that $x \in w \in L - M$, and note that

$w \neq u$ and $w \neq v$. Thus, we have exhibited u, v, and w, comprising three branches of L such that x is an element of each, contradicting the fact that $\deg_L(x) = 2$.

2.4 Branches and Vertices of a Loop

Suppose that L is a loop. It is immediate that L must have more than two branches. A little experimentation in a geometrical realization of the network $(\sigma L, L)$ should convince the reader that the number of branches of L must be equal to the number of vertices of σL. We cite this fact as the following Theorem.

Theorem. *Let L be a loop.*
Then,

    (1) $pL \geq 3$.

    (2) $pL = p\sigma L$.

Proof: Since (1) is easily verified we prove (2) by induction on pL.

If $pL = 3$, it is immediate that $p\sigma L = 3$. Thus, let $m \in \omega$ such that $m \geq 3$ and suppose that $pL = m + 1$. Also, assume that if M is a loop and $pM = m$, then $pM = p\sigma M$. We shall show that $pL = p\sigma L$.

Pick any $x \in \sigma L$. Let y and z be such that $y \neq z$ and $x \in \{x, y\} \in L$ and $x \in \{x, z\} \in L$. Now, if $\{y, z\} \in L$, it is clear that

$$\{\{x, y\}\} \cup \{\{x, z\}\} \cup \{\{y, z\}\} \text{ is a loop,}$$
$$\{\{x, y\}\} \cup \{\{x, z\}\} \cup \{\{y, z\}\} \subset L,$$
$$\{\{x, y\}\} \cup \{\{x, z\}\} \cup \{\{y, z\}\} \neq L,$$

contradicting the Theorem of 2.3. Thus, we must have $\{y, z\} \notin L$.

Now, let
$$M = (L - \{\{x, y\}, \{x, z\}\}) \cup \{\{y, z\}\}.$$

We assert that M is a loop. Since it is immediate that $\deg_M(u) = 2$ for each $u \in \sigma M$, we need only show that M is connected.

Suppose that M is not connected.
Then, there exists K such that K is a component of M and $\{y, z\} \notin K$. It is immediate that $\deg_K(u) = 2$ for each $u \in \sigma K$, and thus, K is a loop. But $K \subset L$ and $K \neq L$, contradicting the Theorem of 2.3. Hence, M is connected and M is a loop.

But $pM = m$.

By the inductive hypothesis, $pM = p\sigma M$.

But since $\sigma M = \sigma L - \{x\}$,

$$p\sigma L = 1 + p\sigma M = 1 + pM = 1 + m = pL.$$

The proof is complete.

It is interesting to note that the converse of the preceding Theorem is false. More precisely, if $(\sigma L, L)$ is such a network that $p\sigma L = pL \geq 3$, it is not true that L is a loop. In fact, even if we add the hypothesis that L is connected, it is not true that L is a loop. As a counterexample consider a network $(\sigma L, L)$ with this geometrical realization.

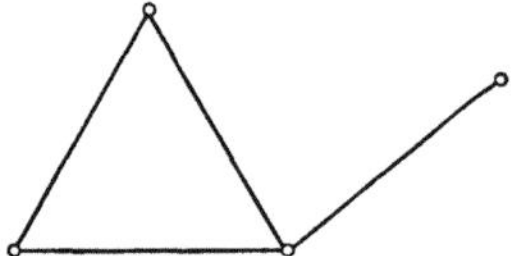

Observe, however, that there exists a subset of L which is a loop. We shall prove that this is always the case whenever $(\sigma L, L)$ is such a network that $p\sigma L = pL \geq 3$.

2.5 Paths in a Loop

Let L be a loop and let $\{x, y\} \in L$. Observe that the trivial path $\{(1, \{x, y\})\}$ is indeed a proper path from x to y in L. However, a look at a geometrical realization of $(\sigma L, L)$ shows that there must exist another proper path from x to y in L which does not include the branch $\{x, y\}$. Thus, this second path is a proper path from x to y in $L - \{\{x, y\}\}$. In the following Theorem 1 we establish the existence of this second proper path from x to y in $L - \{\{x, y\}\}$. Then, in Theorem 2 we point out that this proper path from x to y in $L - \{\{x, y\}\}$ must include every branch of $L - \{\{x, y\}\}$ and that all vertices of the network $(\sigma L, L)$ appear in this proper path.

Theorem 1. *Let L be a loop and let $\{x, y\} \in L$.*
Then,
for some f, f is a proper path from x to y in $L - \{\{x, y\}\}$.

Proof: Suppose no proper path exists from x to y in $L - \{\{x, y\}\}$. Let

$$H = \left\{ z \mid \text{for some } f, f \text{ is a proper path from } x \text{ to } z \text{ in } L - \{\{x, y\}\} \right\}.$$

By our assumption above, $y \notin H$.

Let

$$Y = \left\{ f \mid \text{for some } z, f \text{ is a proper path from } x \text{ to } z \text{ in } L - \{\{x, y\}\} \right\}.$$

Observe trivially that $Y \neq 0$, since for some t, $\{x, t\} \in L$ and $t \neq y$, implying that $\{(1, \{x, t\})\} \in Y$.

We want to consider the maximal number of elements of rng f, for $f \in Y$. Thus, let

$$m = \max \{p \operatorname{rng} f \mid f \in Y\}.$$

Produce $g \in Y$ such that $p \operatorname{rng} g = m$.

In the remainder of the proof, which is divided into two parts, we show that $y \in H$.

Part 1. $m = pL - 1$.

Proof of Part 1: For each $f \in Y$, rng $f \subset L - \{\{x, y\}\}$. Thus, $m \leq p(L - \{\{x, y\}\}) = pL - 1$. Suppose that $m < pL - 1$.

Produce z such that g is a proper path from x to z in $L - \{\{x, y\}\}$. Observe that $z \in H$, implying that $z \neq y$. Also $z \neq x$.

But $z \in \sigma L$, implying that $\deg_L(z) = 2$.

Since $z \neq x$ and $z \neq y$, $\deg_{L - \{\{x, y\}\}}(z) = 2$.

On the other hand, since g is a proper path from x to z in $L - \{\{x, y\}\}$, $\deg_{\operatorname{rng} g}(z) = 1$.

Thus, produce r such that $\{z, r\} \in L - \{\{x, y\}\}$ and $\{z, r\} \notin \operatorname{rng} g$.

We examine this vertex r in detail, noting first that since $r \in \sigma L$, $\deg_L(r) = 2$.

Observe that if $r \in \sigma \operatorname{rng} g - \{x\}$, then $\deg_{\operatorname{rng} g}(r) = 2$, which, together with $\{z, r\} \in L - \operatorname{rng} g$, implies that $\deg_L(r) \geq 3$, which is false.

Thus, $r \notin \sigma \operatorname{rng} g - \{x\}$.

Futhermore if $r = x$, then $\deg_{\operatorname{rng} g}(x) = 1$ and $x \in \{x, y\} \in L - \operatorname{rng} g$ and $x = r \in \{z, r\} \in L - \operatorname{rng} g$, implying that $\deg_L(x) \geq 3$, which is false.

Thus, $r \notin \sigma \operatorname{rng} g$, and in particular, $r \neq x$.

With this established, however, it is easy to show that

$$g \cup \{(m + 1, \{z, r\})\}$$

is a proper path from x to r in $L - \{\{x, y\}\}$ with $p \operatorname{rng} (g \cup \{(m + 1, \{z, r\})\}) = m + 1$, contradicting the maximality of m. This completes the proof of Part 1.

Part 2. $y \in H$.

Proof of Part 2: By (2) of the Theorem of 1.15 and Part 1,

$$p \sigma \operatorname{rng} g = 1 + p \operatorname{rng} g = 1 + (pL - 1) = pL.$$

By (2) of the Theorem of 2.4, $p \sigma \operatorname{rng} g = pL = p \sigma L$.

Since $\sigma \operatorname{rng} g \subset \sigma L$, we conclude that $\sigma \operatorname{rng} g = \sigma L$.

But $y \in \sigma L$.

Thus, $y \in \sigma \operatorname{rng} g$.

Find the first branch of rng g of which y is an element. Specifically, let

$$k = \min \{j \mid j \in [pL - 1] \text{ and } y \in g_j\}.$$

Thus, it is clear that $g \mid [k]$ is a proper path from x to y in $L - \{\{x, y\}\}$, implying that $y \in H$.

Theorem 2. *Let L be a loop. Let $\{x, y\} \in L$. Let g be a proper path from x to y in $L - \{\{x, y\}\}$.*
Then,

$\quad$ (1) $\quad \operatorname{rng} g = L - \{\{x, y\}\}$.

$\quad$ (2) $\quad \sigma \operatorname{rng} g = \sigma L$.

Proof of (1): Note that rng $g \subset L - \{\{x, y\}\}$, implying that rng $g \cup \{\{x, y\}\} \subset L$.
Since g is a proper path from x to y in $L - \{\{x, y\}\}$, it is easy to check that rng $g \cup \{\{x, y\}\}$ is a loop.

By the Theorem of 2.3, rng $g \cup \{\{x, y\}\} = L$, implying that rng $g = L - \{\{x, y\}\}$.

Proof of (2): By (1) $\sigma \operatorname{rng} g = \sigma(L - \{\{x, y\}\})$.
But since $\deg_L(x) = 2$ and $\deg_L(y) = 2$, it is immediate that $\sigma(L - \{\{x, y\}\}) = \sigma L$.

2.6 Removal of a Branch from a Loop

In the last paragraphs we considered a loop L and a branch $\{x, y\} \in L$ and then examined $L - \{\{x, y\}\}$, the set of branches remaining after removing the branch $\{x, y\}$ from the loop L. Recall that we have considered this situation in a more general setting in 1.26, and by Theorem 1 of 1.26, we can assert that the number of components of $L - \{\{x, y\}\}$ does not exceed 2. However, in our special case here when L is a loop we can make the stronger statement that the number of components of $L - \{\{x, y\}\}$ is exactly 1. This is, of course, equivalent to the statement that $L - \{\{x, y\}\}$ is connected, and is the content of the following Theorem.

Theorem. *Let L be a loop. Let $\{x, y\} \in L$.*
Then,
$\quad L - \{\{x, y\}\}$ is connected.

Proof: By Theorem 1 of 2.5, produce f such that f is a proper path from x to y in $L - \{\{x, y\}\}$.

By the Theorem of 1.22, rng f is connected.
By (1) of Theorem 2 of 2.5, rng $f = L - \{\{x, y\}\}$.

2.7 Tree in a Network

Usually in the literature a tree T in a network (V, S) is defined as a connected subset of branches of S such that T includes no loop as a subset. Formally,

T is a tree in (V, S)

if and only if

(1) (V, S) is a network.

(2) $T \subset S$.

(3) T is connected.

(4) $\{L | L \subset T \text{ and } L \text{ is a loop}\} = 0$.

The reader should be cautioned, however, that the above definition is not completely uniform in the literature. Some writers prefer to add the additional requirement that every vertex of the network (V, S) must appear in some branch of T. On our notation this additional condition is equivalent to the statement $\sigma T = V$, but we prefer not to include this condition in our definition.

2.8 Trees

Recall that in 2.1 we defined a loop L in a network (V, S), but then pointed out in 2.2 that the network (V, S) in which the loop was imbedded was of no great significance. In 2.2 we offered an intrinsic definition of a loop L, independent of the network in which the loop was imbedded.

The situation is similar in the case of trees. If T is a tree in a network (V, S), then the network (V, S) is of only minor importance. Thus, we prefer to exhibit an intrinsic definition of a tree. This is accomplished by stating that any set T is a tree if and only if T is a tree in the network $(\sigma T, T)$. Formally,

T is a tree

if and only if

T is a tree in $(\sigma T, T)$.

2.9 Connected Subset of a Tree

Let T be a tree and let M be a connected subset of T. Clearly, M ins
a tree, because if L is a loop which is a subset of M, then L is a subset
of T, which is impossible. We cite this fact as the following Theorem,
but omit the proof.

Theorem. *Let T be a tree. Let $M \subset T$ such that M is connected.*
Then,
  M *is a tree.*

2.10 Branches and Vertices of a Tree

Recall that in 2.4 we compared the number of vertices with the
number of branches of a loop. We want to make the same comparison
in the case of a tree T. A little experimentation in a geometrical realization of the network $(\sigma T, T)$ should convince the reader that the
number of vertices of σT must be one more than the number of branches of T. It turns out that this relation between the number of vertices
and number of branches of a connected set of branches is also sufficient
for the connected set of branches to be a tree. We find it convenient
to separate the propositions citing the necessity and sufficiency of this
relation between the number of vertices and the number of branches
into the following two Theorems. In fact, in the sufficiency proposition
we shall cite and prove the stronger fact that if T is a such connected
set of branches that the number of vertices of σT exceeds the number
of branches of T, then T is a tree.

Theorem 1. *Let T be a tree.*
Then,
  $p\sigma T = 1 + pT.$
  Proof: We prove the Theorem by induction on pT.
  If $pT = 1$, it is immediate that $p\sigma T = 2$.
Thus, let $m \in \omega$ such that $pT = m + 1$.
Also, assume that if M is a tree and $pM \leq m$ then $p\sigma M = 1 + pM$.
We shall show that $p\sigma T = 1 + pT$.
  Select any $\{x, y\} \in T$.
Now, if $\deg_T(x) = 1$ and $\deg_T(y) = 1$, observe that $\{\{x, y\}\}$ is a com-
ponent of T but $\{\{x, y\}\} \neq T$, contradicting the fact that T is connect-
ed. Thus, $\deg_T(x) > 1$ or $\deg_T(y) > 1$.
We assume that $\deg_T(x) > 1$.
  Since $\deg_T(x) > 1$, produce M such that M is a component of
$T - \{\{x, y\}\}$ and $x \in \sigma M$.
By the Theorem of 2.9, M is a tree.
Also, since $M \subset T - \{\{x, y\}\}$, $pM \leq p(T - \{\{x, y\}\}) = m$.

Thus, by the inductive hypothesis, $p\sigma M = 1 + pM$.

We now assert that $y \notin \sigma M$.

For suppose that $y \in \sigma M$.

Then, we can produce f such that f is a proper path from x to y in M.

Observe that $\operatorname{rng} f \cup \{\{x, y\}\}$ is a loop and

$\operatorname{rng} f \cup \{\{x, y\}\} \subset M \cup \{\{x, y\}\} \subset T$, which is false.

Thus, $y \notin \sigma M$.

With these facts established, we let

$$J = \big\{A \,|\, A \text{ is a component of } T - \{\{x, y\}\}\big\},$$

and note that by Theorem 1 of 1.26, $pJ \leq 2$.

We consider two cases, depending on whether $pJ = 1$ or $pJ = 2$.

Case 1. $pJ = 1$.

In this case, $J = \{M\}$ and $M = T - \{\{x, y\}\}$.

Thus, $pM = pT - 1$.

But we have shown that $p\sigma M = 1 + pM$.

Hence $p\sigma M = pT$. On the other hand, since $y \notin \sigma M$, it is clear that $\sigma M = \sigma T - \{y\}$ and $p\sigma M = p\sigma T - 1$.

Thus, $p\sigma T - 1 = p\sigma M = pT$, which is equivalent to $p\sigma T = 1 + pT$, completing Case 1.

Case 2. $pJ = 2$.

In this case, there exists N such that $N \neq M$ and $J = \{M, N\}$.

We have shown that $p\sigma M = 1 + pM$.

The same analysis shows that $p\sigma N = 1 + pN$.

Thus, since $\sigma M \cap \sigma N = 0$,

$p(\sigma M \cup \sigma N) = p\sigma M + p\sigma N = (1 + pM) + (1 + pN) = 2 + pM + pN$.

Now, we have assumed $x \in \sigma M$.

By the Lemma of 1.26, $y \in \sigma N$.

Thus, $\sigma T = \sigma M \cup \sigma N$, and $p\sigma T = p(\sigma M \cup \sigma N) = 2 + pM + pN$.

But $M \cup N = T - \{\{x, y\}\}$ and $M \cap N = 0$. Thus,

$$pT - 1 = p(T - \{\{x, y\}\}) = p(M \cup N) = pM + pN.$$

Combining this with the established fact that $p\sigma T = 2 + pM + pN$ gives $p\sigma T = 1 + pT$, completing the proof.

Theorem 2. *Let T be connected such that $p\sigma T > pT$.*

Then,

　　T is a tree.

Proof: We prove the Theorem by induction on pT.

If $pT = 1$, it is immediate that T is a tree.

Thus, let $m \in \omega$ such that $pT = m + 1$.

Also, assume that if M is connected such that $pM < m$ and $p\sigma M > pM$, then M is a tree.

We shall show that T is a tree.

Suppose that T is not a tree.

Produce L such that $L \subset T$ and L is a loop.

Now if $L = T$, then $\sigma L = \sigma T$, and by (2) of the Theorem of 2.4,

$$pT = pL = p\sigma L = p\sigma T,$$

which is false.

Thus, $L \neq T$.

Produce $\{x, y\}$ such that $\{x, y\} \in T - L$.

Now, if $\deg_T(x) = 1$ and $\deg_T(y) = 1$, observe that $\{\{x, y\}\}$ is a component of T but $\{\{x, y\}\} \neq T$, contradicting the fact that T is connected.

Thus, $\deg_T(x) > 1$ or $\deg_T(y) > 1$.

We assume that $\deg_T(x) > 1$.

Since $\deg_T(x) > 1$, produce M such that M is a component of $T - \{\{x, y\}\}$ and $x \in \sigma M$.

Also let

$$J = \Big\{ A \,\big|\, A \ \text{ is a component of } \ T - \{\{x, y\}\} \Big\},$$

and note that by Theorem 1 of 1.26, $pJ \leq 2$.

We consider two cases, depending on whether $pJ = 1$ or $pJ = 2$, and show that each case is impossible.

Case 1. $pJ = 1$.

In this case $J = \{M\}$ and $M = T - \{\{x, y\}\}$.

Thus, $pM = pT - 1$.

But since $x \in \sigma M$, $\sigma T - \{y\} \subset \sigma M$, implying that
$p\sigma M \geq p(\sigma T - \{y\}) = p\sigma T - 1$.

Hence

$$p\sigma M - pM \geq (p\sigma T - 1) - (pT - 1) = p\sigma T - pT > 0,$$

implying that $p\sigma M > pM$.

But we know $pM = pT - 1 = m$.

Thus, by the inductive hypothesis, M is a tree.

On the other hand $\{x, y\} \in T - L$, implying that $L \subset T - \{\{x, y\}\} = M$.

This contradicts the fact that M is a tree, completing Case 1.

Case 2. $pJ = 2$.

In this case, there exists N such that $N \neq M$ and $J = \{M, N\}$.

By the Lemma of 1.26, $y \in \sigma N$.

Since $x \in \sigma M$, $\sigma T = \sigma M \cup \sigma N$, and thus, since $\sigma M \cap \sigma N = 0$,

$$p\sigma T = p(\sigma M \cup \sigma N) = p\sigma M + p\sigma N.$$

Also, $T - \{\{x, y\}\} = M \cup N$, and, since $M \cap N = 0$,

$$pT - 1 = p(T - \{\{x, y\}\}) = p(M \cup N) = pM + pN.$$

Thus, our initial assumption that $p\sigma T > pT$ yields

$$p\sigma M + p\sigma N > pM + pN + 1.$$

But $M \subset T - \{\{x, y\}\}$, implying that $pM \leq p(T - \{\{x, y\}\}) = pT - 1 = m$.
Also, M is connected.
Now, if $p\sigma M > pM$, the inductive hypothesis would imply that M
is a tree, contradicting the fact that $L \subset M$.
Thus, we must have $pM \geq p\sigma M$, and

$$p\sigma M + p\sigma N > pM + pN + 1 \geq p\sigma M + pN + 1,$$

implying that $p\sigma N > pN + 1$.
In particular, $p\sigma N > pN$.
But N is connected, and since
$N \subset T - \{\{x, y\}\}$, $pN \leq p(T - \{\{x, y\}\}) = pT - 1 = m$. By the inductive hypothesis, N is a tree.
But then, by Theorem 1, $p\sigma N = 1 + pN$.
This contradicts the established fact that $p\sigma N > pN + 1$.
The proof is complete.

2.11 Number of Vertices of a Connected Set of Branches

We can now assess the number of vertices of a connected set of
branches T. If T is a tree, then by Theorem 1 of 2.10, $p\sigma T = 1 + pT$.
On the other hand, if T is not a tree, then we must have $p\sigma T \leq pT$,
for otherwise Theorem 2 of 2.10 is contradicted. Thus, the number
of vertices of σT can never exceed one more than the number of branches
of T, and the upper bound is achieved if and only if T is a tree. We
state these facts as the following Theorem, but omit the proof.

Theorem. *Let T be connected.*
Then,

    (1) $p\sigma T \leq 1 + pT.$

    (2) $p\sigma T = 1 + pT$ *if and only if T is a tree.*

2.12 Addition of a Branch to a Tree

Let T be a tree, and let x and y be distinct vertices of σT such that the branch $\{x, y\}$ is not a branch of T. Consider $T \cup \{\{x, y\}\}$, the connected set of branches formed by adding the branch $\{x, y\}$ to the tree T. Observe that the vertices of $\sigma(T \cup \{\{x, y\}\})$ are precisely the vertices of σT, but $T \cup \{\{x, y\}\}$ has one more branch than T. Thus, since T is a tree, and $p\sigma T = 1 + pT$, we see that number of vertices of $\sigma(T \cup \{\{x, y\}\})$ is equal to the number of branches of $T \cup \{\{x, y\}\}$. It is impossible for $T \cup \{\{x, y\}\}$ to be a tree, and there must exist L such that L is a loop and $L \subset T \cup \{\{x, y\}\}$. Furthermore $\{x, y\} \in L$, for otherwise $L \subset T$, contradicting the fact that T is a tree. Finally, it is possible to show that such a loop $L \subset T \cup \{\{x, y\}\}$ is unique. These facts are cited in the following Theorem.

Theorem. *Let T be a tree. Let $x \in \sigma T$ and let $y \in \sigma T$ such that $x \neq y$. Suppose that $\{x, y\} \notin T$.*
Then,

$\quad p\left\{L \,|\, L \text{ is a loop and } \{x, y\} \in L \subset T \cup \{\{x, y\}\}\right\} = 1.$

Proof: Consider $T \cup \{\{x, y\}\}$ and note that $T \cup \{\{x, y\}\}$ is connected. Since $\{x, y\} \notin T$, $p(T \cup \{\{x, y\}\}) = 1 + pT$.
Thus, since T is a tree, $p(T \cup \{\{x, y\}\}) = 1 + pT = p\sigma T$.

On the other hand, since $x \in \sigma T$ and $y \in \sigma T$, $\sigma(T \cup \{\{x, y\}\}) = \sigma T$, and $p\sigma(T \cup \{\{x, y\}\}) = p\sigma T$.
Thus, $p(T \cup \{\{x, y\}\}) = p\sigma(T \cup \{\{x, y\}\})$, and $T \cup \{\{x, y\}\}$ is not a tree. Produce L such that L is a loop and $L \subset T \cup \{\{x, y\}\}$.
Note that $\{x, y\} \in L$, for otherwise $L \subset T$, contradicting the fact that T is a tree.

To complete the proof, suppose that M is a loop such that $M \neq L$ and $\{x, y\} \in M \subset T \cup \{\{x, y\}\}$.
We shall show that this leads to a contradiction.

In particular, in the remainder of the proof which is divided into three Parts, we shall show in Part 2 that $p\sigma(L \cap M) \leq p(L \cap M)$, while in Part 3 we shall show that $p\sigma(L \cap M) > p(L \cap M)$.

Part 1. $p(\sigma L \cup \sigma M) = p(L \cup M)$.

Proof of Part 1: By the Theorem of 2.6, $L - \{\{x, y\}\}$ is connected and $M - \{\{x, y\}\}$ is connected.
Since $x \in \sigma(L - \{\{x, y\}\}) \cap \sigma(M - \{\{x, y\}\})$, the Theorem of 1.21 guarantees that $(L - \{\{x, y\}\}) \cup (M - \{\{x, y\}\})$ is connected.
But since

$$(L - \{\{x, y\}\}) \cup (M - \{\{x, y\}\}) = (L \cup M) - \{\{x, y\}\},$$

we see that $(L \cup M) - \{\{x, y\}\}$ is connected and $(L \cup M) - \{\{x, y\}\} \subset T$.

By the Theorem of 2.9, $(L \cup M) - \{\{x, y\}\}$ is a tree, and

$$p\sigma\big((L \cup M) - \{\{x, y\}\}\big) = 1 + p\big((L \cup M) - \{\{x, y\}\}\big).$$

But since $x \in \sigma(L \cup M)$ and $y \in \sigma(L \cup M)$,
$\sigma((L \cup M) - \{\{x, y\}\}) = \sigma(L \cup M)$. Also, since
$\{x, y\} \in L \cup M$, $p((L \cup M) - \{\{x, y\}\}) = p(L \cup M) - 1$. Thus, we have

$$p\sigma(L \cup M) = p\sigma((L \cup M) - \{\{x, y\}\}) = 1 + p\big((L \cup M) - \{\{x, y\}\}\big)$$
$$= 1 + (p(L \cup M) - 1) = p(L \cup M).$$

To complete the proof of Part 1, note trivially that $\sigma(L \cup M) = \sigma L \cup \sigma M$.

Part 2. $p\sigma(L \cap M) \le p(L \cap M)$.

Proof of Part 2: For any two finite sets A and B, note that

$$p(A \cup B) = pA + pB - p(A \cap B).$$

Applying this fact twice to the results of Part 1,

$$p\sigma L + p\sigma M - p(\sigma L \cap \sigma M) = pL + pM - p(L \cap M).$$

By (2) of the Theorem of 2.4, $p\sigma L = pL$ and $p\sigma M = pM$.
Thus, $p(\sigma L \cap \sigma M) = p(L \cap M)$.
However, check trivally that $\sigma(L \cap M) \subset \sigma L \cap \sigma M$.
Thus, $p\sigma(L \cap M) \le p(\sigma L \cap \sigma M) = p(L \cap M)$, completing the proof of
Part 2.

Part 3. $p\sigma(L \cap M) > p(L \cap M)$.

Proof of Part 3: Let

$$J = \{A \,|\, A \text{ is a component of } L \cap M\}.$$

Since $\{x, y\} \in L \cap M \neq 0$, $(\sigma(L \cap M),\ L \cap M)$ is a network, and by
(2) of the Theorem of 1.25,

$$\sigma(L \cap M) = \sigma\sigma J \cup \{u \,|\, \deg_{L \cap M}(u) = 0\}.$$

But $\sigma\sigma J \cap \{u \,|\, \deg_{L \cap M}(u) = 0\} = 0$, and,

$$p\sigma(L \cap M) = p(\sigma\sigma J \cup \{u \,|\, \deg_{L \cap M}(u) = 0\})$$
$$= p\sigma\sigma J + p\{u \,|\, \deg_{L \cap M}(u) = 0\} \ge p\sigma\sigma J.$$

Thus, $p\sigma(L \cap M) \ge p\sigma\sigma J$.
But $\sigma\sigma J = \sigma\sigma\{A \,|\, A \in J\} = \sigma\{\sigma A \,|\, A \in J\}$.

Furthermore, if $A \in J$ and $A' \in J$ with $A \neq A'$, then $\sigma A \cap \sigma A' = 0$.
Thus,

$$p\sigma(L \cap M) \geq p\sigma\sigma J = p\sigma\{\sigma A | A \in J\} = \sum_{A \in J} p\sigma A.$$

But consider any $A \in J$.
Note that A is connected and $A \subset L \cap M \subset L$.
By the Theorem of 2.3, it is not the case that $L \subset M$.
Thus, $L \cap M \neq L$, and $A \neq L$.
If A is not a tree there exists a loop N such that $N \subset A \subset L$ and $N \neq L$,
contradicting the Theorem of 2.3.
Thus, A is a tree and $p\sigma A = 1 + pA$.

Using the fact that $p\sigma A = 1 + pA$, for each $A \in J$,

$$p\sigma(L \cap M) \geq \sum_{A \in J} p\sigma A = \sum_{A \in J} (1 + pA) = pJ + \sum_{A \in J} pA.$$

But since $L \cap M \neq 0$, $J \neq 0$ and $pJ > 0$.
Thus

$$p\sigma(L \cap M) \geq pJ + \sum_{A \in J} pA > \sum_{A \in J} pA.$$

But if $A \in J$ and $A' \in J$ with $A \neq A'$, then $A \cap A' = 0$.
Thus,

$$p\sigma(L \cap M) > \sum_{A \in J} pA = p\sigma\{A | A \in J\} = p\sigma J.$$

By (1) of the Theorem of 1.25, $L \cap M = \sigma J$, and the proof of Part 3
is complete.

2.13 Existence of Maximal Trees

Let (V, S) be a connected network with S non-empty. Since a set
consisting of a single branch of S is clearly a tree, there exist trees in
(V, S). We ask whether there exists a tree in (V, S) whose vertices
comprise all of the vertices of V. This question is answered affirmatively
in the following Theorem. Such a tree in (V, S) whose vertices comprise
all of the vertices of V is called by some writers a maximal tree in
(V, S), and these maximal trees will frequently be used in our later
work.

Theorem. *Let (V, S) be a connected network. Suppose that $S \neq 0$.*
Then,

for some T,

(1)   *T is a tree in (V, S).*

(2)   *$\sigma T = V$.*

Proof: Let

$$K = \{A \,|\, A \text{ is a tree in } (V, S)\}.$$

Since $S \neq 0$, there exists an element of K consisting of a single branch of S, and thus $K \neq 0$.
Let

$$m = \max \{p\sigma A \,|\, A \in K\}.$$

Produce Q such that $Q \in K$ and $p\sigma Q = m$.
We shall show that $\sigma Q = V$.

Now, since $Q \subset S$, $\sigma Q \subset \sigma S = V$.
Suppose that $\sigma Q \neq V$.
Produce x such that $x \in V - \sigma Q$.
Also, produce y such that $y \in \sigma Q$.

Since (V, S) is connected, produce f such that f is a path from y to x in S.
Observe that $x \in f_{pf} \cap (V - \sigma Q)$.
Let

$$i = \min \{j \,|\, j \in \operatorname{dmn} f \text{ and } f_j \cap (V - \sigma Q) \neq 0\}.$$

Consider f_i, and let $f_i = \{r, s\} \in S$.

Now, if $r \in V - \sigma Q$ and $s \in V - \sigma Q$, then the minimality of i is contradicted.
Thus, assume that $s \in V - \sigma Q$ and $r \notin V - \sigma Q$.
Hence, $\{r, s\} \notin Q$ and $r \in \sigma Q$.

Consider $Q \cup \{\{r, s\}\}$.
Clearly, since $r \in \sigma Q$, $Q \cup \{\{r, s\}\}$ is connected.
Also, since Q is a tree and $\{r, s\} \notin Q$.

$$\begin{aligned}
p\sigma(Q \cup \{\{r, s\}\}) &= p(\sigma Q \cup \{r, s\}) \\
&= p(\sigma Q \cup \{s\}) \\
&= p\sigma Q + 1 \\
&= (pQ + 1) + 1 \\
&= p(Q \cup \{\{r, s\}\}) + 1.
\end{aligned}$$

Thus, $Q \cup \{\{r, s\}\}$ is a tree in (V, S).
But from above, $p\sigma(Q \cup \{\{r, s\}\}) = p\sigma Q + 1 = m + 1$, contradicting the maximality of m.
The proof is complete.

2.14 Cut Set in a Network

The final special set of branches we consider in this Chapter is a cut set in a network. A cut set in a network (V, S) is a subset of S whose removal from S retaining the vertices of V leaves a disconnected subnetwork of (V, S), but such that the cut set is a minimal set of branches with this property. Thus, if M is a cut set in (V, S), then M is such a subset of S that the subnetwork $(V, S - M)$ is disconnected, but the subnetwork $(V, S - (M - \{\{x, y\}\}))$ is connected for each branch $\{x, y\}$ of M. Formally,

> M is a cutset in (V, S)
>
> if and only if
>
> (1) (V, S) is a network.
> (2) $M \subset S$.
> (3) $(V, S - M)$ is not a connected network.
> (4) $\{x, y\} \in M$
>   implies
> $(V, S - (M - \{\{x, y\}\}))$ is a connected network.

Clearly, if (V, S) is a not a connected network, then the empty set is a cut set in (V, S). Thus, a cut set in a network (V, S) will be of interest to us only in the special case when (V, S) is connected.

2.15 Existence of Cut Sets

Consider a connected network (V, S) with S non-empty. Although we have defined the meaning of a cut set in (V, S), we have yet to demonstrate the existence of a cut set in (V, S). In this Paragraph we guarantee the existence of a cut set in (V, S). Furthermore, we demonstrate this existence by manufacturing the cut set by means of a process which provides us with a precise description of those branches comprising the cut set.

Our manufacturing process will require the use of a tree in (V, S) such that the vertices of T comprise all of the vertices of V. The existence of such a tree is guaranteed by the Theorem of 2.13. Consider any branch $\{s, t\}$ of such a tree T. We shall obtain a cut set M which includes the branch $\{s, t\}$ of T, but will include no other branches of T. Also, a branch $\{x, y\} \in S - T$ will be included in the cut set if and only if the branch $\{s, t\}$ is included in the unique loop including $\{x, y\}$ and contained in $T \cup \{\{x, y\}\}$, as described in the Theorem of 2.12. These facts are cited in the following Theorem.

Theorem. *Let (V, S) be a connected network. Let T be a tree in (V, S). Suppose that $\sigma T = V$. Let $\{s, t\} \in T$. Let L be the function with domain $S - T$ such that for each $\{x, y\} \in S - T$, $L(\{x, y\})$ is the loop such that*

$$\{x, y\} \in L(\{x, y\}) \subset T \cup \{\{x, y\}\}.$$

Let

$$M = \{\{s, t\}\} \cup \{\{x, y\} \mid \{x, y\} \in S - T \text{ and } \{s, t\} \in L(\{x, y\})\}.$$

Then,

(1) $M \cap T = \{\{s, t\}\}.$

(2)   *M is a cut set in* $(V, S).$

Proof: Since (1) is immediate from the definition of M, we need only prove (2). The proof of (2) is divided into two parts.

Part 1. $(V, S - M)$ is not a connected network.

Proof of Part 1: If, for some $x \in V$, $\deg_{S-M}(x) = 0$, then $(V, S - M)$ is not a connected network and we are through.

Thus, we assume that for each $x \in V$, $\deg_{S-M}(x) > 0$.

By Theorem 3 of 1.18, $V = \sigma(S - M)$.

Hence, it suffices to show that $S - M$ is not a connected set of branches.

We apply the Theorem of 1.20 and show that $S - M$ is not a path connected set of branches.

In particular, we show that there does not exist a proper path from s to t in $S - M$.

Suppose that f is a proper path from s to t in $S - M$.

Let $m \in \omega$ such that $\operatorname{dmn} f = [m]$.

By the Theorem of 1.15, produce a univalent sequence b with $\operatorname{dmn} b = [m + 1]$ such that

$$\begin{aligned}
b_1 &= s, \\
b_{m+1} &= t, \\
f_i &= \{b_i, b_{i+1}\} \text{ for each } i \in [m].
\end{aligned}$$

Now, if $m = 1$, then $\operatorname{rng} f = \{\{b_1, b_{m+1}\}\} = \{\{s, t\}\} \subset M$, contradicting the fact that f is a path in $S - M$.

Thus, we may assume that $m > 1$.

For each $i \in [m]$, let $P(i)$ be the following proposition:

> There exists a proper path from s to b_{i+1}
> in $T - \{\{s, t\}\}.$

We shall show, by induction, that $P(i)$ is true for each $i \in [m]$.

To establish $P(1)$ we construct a proper path from s to b_2 in $T - \{\{s, t\}\}$.

Consider the branch $\{s, b_2\} \in S$.

Since $m > 1$, we know that $b_2 \neq t$.

Now, if $\{s, b_2\} \in T$, then the function $\{(1, \{s, b_2\})\}$ is trivially a proper path from s to b_2 in $T - \{\{s, t\}\}$.

Thus, we may assume that $\{s, b_2\} \in S - T$.

4*

Consider $L(\{s, b_2\})$ which is a loop such that

$$\{s, b_2\} \in L(\{s, b_2\}) \subset T \cup \{\{s, b_2\}\}.$$

By Theorem 1 of 2.5, produce z such that z is a proper path from s to b_2 in $L(\{s, b_2\}) - \{\{s, b_2\}\}$.

The construction of $L(\{s, b_2\})$ guarantees that z is a proper path from s to b_2 in T.

But $\{s, b_2\} \in \text{rng } f \subset S - M$, implying that $\{s, b_2\} \notin M$.

Thus, $\{s, t\} \notin L(\{s, b_2\})$, implying that z is a proper path from s to b_2 in $T - \{\{s, t\}\}$.

This completes the proof that $P(1)$ is true:

Next, suppose that $i \in [m]$ and $i + 1 \in [m]$ such that $P(i)$ is true. We shall show that $P(i + 1)$ is true by exhibiting a proper path from s to b_{i+2} in $T - \{\{s, t\}\}$.

Since $P(i)$ is true, produce g such that g is a proper path from s to b_{i+1} in $T - \{\{s, t\}\}$.

Let $k \in \omega$ such that $\text{dmn } g = [k]$.

Consider the branch $\{b_{i+1}, b_{i+2}\} \in S$,

Clearly, $\{b_{i+1}, b_{i+2}\} \neq \{s, t\}$.

Now, if $\{b_{i+1}, b_{i+2}\} \in T$, then the function

$$g \cup \{(k + 1, \{b_{i+1}, b_{i+2}\})\}$$

is trivially a proper path from s to b_{i+2} in $T - \{\{s, t\}\}$ and we are through.

Thus, we may assume that $\{b_{i+1}, b_{i+2}\} \in S - T$.

Consider $L(\{b_{i+1}, b_{i+2}\})$ which is a loop such that

$$\{b_{i+1}, b_{i+2}\} \in L(\{b_{i+1}, b_{i+2}\}) \subset T \cup \{\{b_{i+1}, b_{i+2}\}\}.$$

By Theorem 1 of 2.5, produce w such that w is a proper path from b_{i+1} to b_{i+2} in $L(\{b_{i+1}, b_{i+2}\}) - \{\{b_{i+1}, b_{i+2}\}\}$.

The construction of $L(\{b_{i+1}, b_{i+2}\})$ guarantees that w is a proper path from b_{i+1} to b_{i+2} in T.

But $\{b_{i+1}, b_{i+2}\} \in \text{rng } f \subset S - M$, implying that $\{b_{i+1}, b_{i+2}\} \notin M$. Thus, $\{s, t\} \notin L(\{b_{i+1}, b_{i+2}\})$, implying that w is a proper path from b_{i+1} to b_{i+2} in $T - \{\{s, t\}\}$.

Recapitulating, we have g such that g is a proper path from s to b_{i+1} in $T - \{\{s, t\}\}$, and we have w such that w is a proper path from b_{i+1} to b_{i+2} in $T - \{\{s, t\}\}$. From these two proper paths it is easy to manufacture a path from s to b_{i+2} in $T - \{\{s, t\}\}$, and Theorem 1.16 applied to this path from s to b_{i+2} in $T - \{\{s, t\}\}$ yields a proper path from s to b_{i+2} in $T - \{\{s, t\}\}$. Thus, $P(i + 1)$ has been established.

Since $P(i)$ is true for each $i \in [m]$, consider $P(m)$, noting that $b_{m+1} = t$, and produce h such that h is a proper path from s to t in $T - \{\{s, t\}\}$.

Then, it is clear that rng $h \cup \{\{s, t\}\}$ is a loop and rng $h \cup \{\{s, t\}\} \subset T$, which is impossible.

This completes the proof of Part 1.

Part 2. $\{x, y\} \in M$ implies $\big(V, S - (M - \{\{x, y\}\})\big)$ is a connected network.

Proof of Part 2: Let $\{x, y\} \in M$.

Noting the Theorem of 1.20, it suffices to show that
$V = \sigma\big(S - (M - \{\{x, y\}\})\big)$ and $S - (M - \{\{x, y\}\})$ is a path connected set of branches.

If we can show that whenever $c \in V$ and $d \in V$, with $c \neq d$, then there exists a path from c to d in $S - (M - \{\{x, y\}\})$, we will have shown that $V = \sigma\big(S - (M - \{\{x, y\}\})\big)$ and $S - (M - \{\{x, y\}\})$ is a path connected set of branches.

Thus, let $c \in V$ and let $d \in V$ with $c \neq d$.
We shall exhibit a path from c to d in $S - (M - \{\{x, y\}\})$.

Since T is connected, produce f such that f is a proper path from c to d in T.
Let $m \in \omega$ such that dmn $f = [m]$.
By the Theorem of 1.15, produce a univalent sequence b with dmn $b = [m + 1]$ such that

$$
\begin{aligned}
b_1 &= c, \\
b_{m+1} &= d, \\
f_i &= \{b_i, b_{i+1}\} \text{ for each } i \in [m].
\end{aligned}
$$

Now, if $\{s, t\} \notin$ rng f, then, noting that $M \cap T = \{\{s, t\}\}$,

$$
\text{rng } f \subset T - \{\{s, t\}\} \subset S - M \subset S - (M - \{\{x, y\}\}),
$$

and we are through.
Thus, we may assume that $\{s, t\} \in$ rng f.
We lose no generality by assuming that for some $k \in [m]$, $s = b_k$ and $t = b_{k+1}$ and $\{s, t\} = \{b_k, b_{k+1}\} = f_k$.

Next, suppose that $\{x, y\} = \{s, t\}$.
In this case, it is easy to check that

$$
\text{rng } f \subset T \subset \{\{s, t\}\} \cup (S - M) = S - (M - \{\{s, t\}\}),
$$

and since $\{x, y\} = \{s, t\}$, we are through.
Thus, we assume that $\{x, y\} \neq \{s, t\}$.
Hence, $\{s, t\} \in L(\{x, y\})$.

By Theorem 1 of 2.5, produce g such that g is a proper path from s to t in $L(\{x, y\}) - \{\{s, t\}\}$.

Since $L(\{x, y\}) \subset T \cup \{\{x, y\}\}$, we know that g is a proper path from s to t in $(T \cup \{\{x, y\}\}) - \{\{s, t\}\}$.

Now, if $c = s$, then g is a proper path from c to t in $(T \cup \{\{x, y\}\}) - \{\{s, t\}\}$.

On the other hand, if $c \neq s$, then $k > 1$ and $f|[k-1]$ is a proper path from c to s in $T - \{\{s, t\}\}$.

Thus, by combining $f|[k-1]$ with g, which is a proper path from s to t in $(T \cup \{\{x, y\}\}) - \{\{s, t\}\}$, we can easily produce a path from c to t in $(T \cup \{\{x, y\}\}) - \{\{s, t\}\}$.

Hence, in any case, we can produce h such that h is a path from c to t in $(T \cup \{\{x, y\}\}) - \{\{s, t\}\}$.

Now, if $t = d$, then h is a path from c to d in $(T \cup \{\{x, y\}\}) - \{\{s, t\}\}$. On the other hand, if $t \neq d$, note that

$$\{(i, f_{k+i}) \,|\, i \in [m - k]\}$$

is a proper path from t to d in $T - \{\{s, t\}\}$.

Thus, by combining $\{(i, f_{k+i}) \,|\, i \in [m - k]\}$ with h, which is a path from c to t in $(T \cup \{\{x, y\}\}) - \{\{s, t\}\}$, we can easily produce a path from c to d in $(T \cup \{\{x, y\}\}) - \{\{s, t\}\}$.

Hence, in any case, we can produce u such that u is a path from c to d in $(T \cup \{\{x, y\}\}) - \{\{s, t\}\}$.

But since $T \cap M = \{\{s, t\}\}$, it is easy to check that

$$\mathrm{rng}\ u \subset (T \cup \{\{x, y\}\}) - \{\{s, t\}\} \subset S - (M - \{\{x, y\}\}).$$

Thus, u is a path from c to d in $S - (M - \{\{x, y\}\})$, completing the proof.

2.16 Alternate Characterization of Cut Sets

If (V, S) is a connected network with S non-empty, and if T is a tree in (V, S) such that the vertices of T comprise all of the vertices of V, and if $\{s, t\}$ is a branch of T, we described in 2.15 a certain cut set M in (V, S) including the branch $\{s, t\}$. In particular, the cut set M included no other branches of T, and included those branches $\{x, y\} \in S - T$ such that the branch $\{s, t\}$ is included in the unique loop including $\{x, y\}$ and contained in $T \cup \{\{x, y\}\}$. In this Paragraph we show that the same cut set can be characterized in another way which will be more convenient for our later work.

To describe this alternate characterization of the cut set we must introduce the concept of the star of a vertex with respect to a set of branches. In particular, if (V, S) is any network and if x is a vertex of V

and if $L \subset S$, then the star of x in V, with respect to L, denoted by $\text{star}_L^V(x)$, is defined to be that set of vertices of V consisting of x plus all other vertices of V which can be joined to x by a path in L. Formally,

$$\text{star}_L^V(x) = \{x\} \cup \{y \mid y \in V \text{ and for some } f, f \text{ is a path from } x \text{ to } y \text{ in } L\}.$$

Now, let (V, S) be a connected network with S non-empty, and let T be a tree in (V, S) such that the vertices of T comprise all of the vertices of V, and let $\{s, t\}$ be a branch of T. In the following Theorem we show that the above cut set M in (V, S) can be simply described as the set of all branches of S such that one vertex of the branch is an element of $\text{star}_{T-\{\{s,t\}\}}^V(s)$ while the other vertex of the branch is not an element of $\text{star}_{T-\{\{s,t\}\}}^V(s)$.

Theorem. *Let (V, S) be a connected network. Let T be a tree in (V, S). Suppose that $\sigma T = V$. Let $\{s, t\} \in T$. Let L be that function with domain $S - T$ such that for each $\{x, y\} \in S - T$, $L(\{x, y\})$ is the loop such that*

$$\{x, y\} \in L(\{x, y\}) \subset T \cup \{\{x, y\}\}.$$

Let

$$M = \{\{s, t\}\} \cup \{\{x, y\} \mid \{x, y\} \in S - T \text{ and } \{s, t\} \in L(\{x, y\})\}.$$

Let

$$P = S \cap \{\{x, y\} \mid x \in \text{star}_{T-\{\{s,t\}\}}^V(s) \text{ and } y \notin \text{star}_{T-\{\{s,t\}\}}^V(s)\}.$$

Then,
    $M = P$.

Proof:
Part 1. $M \subset P$.
Proof of Part 1: Let $\{x, y\} \in M$.

Suppose first that $\{x, y\} = \{s, t\}$.
Now, $s \in \text{star}_{T-\{\{s,t\}\}}^V(s)$.
Thus, it suffices to show that $t \notin \text{star}_{T-\{\{s,t\}\}}^V(s)$.

But if $t \in \text{star}_{T-\{\{s,t\}\}}^V(s)$, then there exists f such that f is a proper path from s to t in $T - \{\{s, t\}\}$. Clearly, $\text{rng } f \cup \{\{s, t\}\}$ is a loop and $\text{rng } f \cup \{\{s, t\}\} \subset T$, which is impossible.

Next, we suppose $\{x, y\} \neq \{s, t\}$.
Hence, $\{s, t\} \in L(\{x, y\})$.

By Theorem 1 of 2.5, produce g such that g is a proper path from s to t in $L(\{x, y\}) - \{\{s, t\}\}$.
Let $m \in \omega$ such that $\text{dmn } g = [m]$.

By the Theorem of 1.15, produce a univalent sequence b with $\text{dmn } b = [m + 1]$ such that

$$b_1 = s,$$
$$b_{m+1} = t,$$
$$g_i = \{b_i, b_{i+1}\} \text{ for each } i \in [m].$$

By Theorem 2, (1), of 2.5, $\operatorname{rng} g = L(\{x, y\}) - \{\{s, t\}\}$.
Thus, since $\{x, y\} \in L(\{x, y\})$ and $\{x, y\} \neq \{s, t\}$, we see that $\{x, y\} \in \operatorname{rng} g$.
We lose no generality by assuming that for some $k \in [m]$, $x = b_k$ and $y = b_{k+1}$ and $\{x, y\} = \{b_k, b_{k+1}\} = g_k$.

Now, if $x = s$, then $x \in \operatorname{star}^V_{T-\{\{s, t\}\}}(s)$.
On the other hand, if $x \neq s$, then $k > 1$, and $g|[k-1]$ is a path form s to x in $T - \{\{s, t\}\}$, implying that $x \in \operatorname{star}^V_{T-\{\{s, t\}\}}(s)$.
Thus, in any case, $x \in \operatorname{star}^V_{T-\{\{s, t\}\}}(s)$.

It suffices to show that $y \notin \operatorname{star}^V_{T-\{\{s, t\}\}}(s)$.
Suppose that $y \in \operatorname{star}^V_{T-\{\{s, t\}\}}(s)$.

Produce w such that w is a proper path from y to s in $T - \{\{s, t\}\}$.
It is easy to check that

$$\{\{s, t\}\} \cup \operatorname{rng} w \cup \operatorname{rng}\left(g|([m] - [k])\right)$$

is a loop and

$$\{\{s, t\}\} \cup \operatorname{rng} w \cup \operatorname{rng}\left(g|([m] - [k])\right) \subset T,$$

which is impossible.

Part 2. $P \subset M$.

Proof of Part 2: Let $\{x, y\} \in P$.

Suppose that $x \in \operatorname{star}^V_{T-\{\{s, t\}\}}(s)$ and $y \notin \operatorname{star}^V_{T-\{\{s, t\}\}}(s)$.
To show that $\{x, y\} \in M$, we assume that $\{x, y\} \neq \{s, t\}$.
Thus, $\{x, y\} \in S - T$.
It suffices to show that $\{s, t\} \in L(\{x, y\})$.

Suppose $\{s, t\} \notin L(\{x, y\})$.
By Theorem 1 of 2.5, produce h such that h is a proper path from y to x in $L(\{x, y\}) - \{\{x, y\}\}$.
Since $L(\{x, y\}) \subset T \cup \{\{x, y\}\}$ and $\{s, t\} \notin L(\{x, y\})$, we conclude that h is a proper path from y to x in $T - \{\{s, t\}\}$.

Since $x \in \operatorname{star}^V_{T-\{\{s, t\}\}}(s)$, produce v such that v is a path from x to s in $T - \{\{s, t\}\}$.
By combining v with h, which is a proper path from y to x in $T - \{\{s, t\}\}$, it is easy to produce a path from y to s in $T - \{\{s, t\}\}$.
But this contradicts $y \notin \operatorname{star}^V_{T-\{\{s, t\}\}}(s)$.
The proof is complete.

Again, let (V, S), T, $\{s, t\}$, and M be as in the preceding Theorem.
We showed in the preceding Theorem that M consists of those branches

of S such that one vertex of the branch is an element of $\text{star}^V_{T-\{\{s,t\}\}}(s)$ and the other vertex of the branch is not an element of $\text{star}^V_{T-\{\{s,t\}\}}(s)$. However, since s and t occur symmetrically in the definition of M, the same Theorem also states that M consists of those branches of S such that one vertex of the branch is an element of $\text{star}^V_{T-\{\{s,t\}\}}(t)$ and the other vertex of the branch is not an element of $\text{star}^V_{T-\{\{s,t\}\}}(t)$.

Thus, as a Corollary, we point out, omitting the proof that those branches of S with one vertex of the branch an element of $\text{star}^V_{T-\{\{s,t\}\}}(s)$ and the other vertex of the branch not an element of $\text{star}^V_{T-\{\{s,t\}\}}(s)$ are precisely the same as those branches of S with one vertex of the branch an element of $\text{star}^V_{T-\{\{s,t\}\}}(t)$ and the other vertex of the branch not an element of $\text{star}^V_{T-\{\{s,t\}\}}(t)$.

Corollary. *Let (V, S) be a connected network. Let T be a tree in (V, S). Suppose that $\sigma T = V$. Let $\{s, t\} \in T$. Let*

$$P = S \cap \{\{x, y\} \mid x \in \text{star}^V_{T-\{\{s,t\}\}}(s) \text{ and } y \notin \text{star}^V_{T-\{\{s,t\}\}}(s)\}.$$

$$Q = S \cap \{\{x, y\} \mid x \in \text{star}^V_{T-\{\{s,t\}\}}(t) \text{ and } y \notin \text{star}^V_{T-\{\{s,t\}\}}(t)\}.$$

Then,

$$P = Q.$$

CHAPTER THREE

Incidence Functions and Incidence Matrices

3.0 Introduction

We now introduce into our structure the concept of the direction of a branch of a network. This is accomplished by means of a function called an incidence function of a set of branches. This incidence function leads to a matrix called an incidence matrix of a set of branches. Most of this Chapter is devoted to properties of such incidence matrices.

3.1 Incidence Functions

Consider a set of branches S and let $\{x, y\} \in S$. Recall that $\{x, y\}$ is merely the set consisting of the elements x and y; it is meaningless to discuss the first element of $\{x, y\}$ or the second element of $\{x, y\}$. Thus, in a geometrical realization of the network $(\sigma S, S)$ the branch joining the vertices corresponding to x and y has no direction. A direction can be assigned to this branch by assigning an order to the elements of the set $\{x, y\}$. This order can be assigned to the elements of the set $\{x, y\}$ by associating with the set $\{x, y\}$ exactly one of the ordered pairs (x, y) or (y, x).

To assign an order to every branch of S, we specify a function f such that $\operatorname{dmn} f = S$, and such that for each $\{x, y\} \in S$, $f(\{x, y\}) = (x, y)$ or $f(\{x, y\}) = (y, x)$.

Such an ordering of every branch of S is called an incidence function of S. Formally,

> f is an incidence function of S
>
> if and only of
>
> (1) $(\sigma S, S)$ is a network.
> (2) f is a function.
> (3) $\operatorname{dmn} f = S$.
> (f) $\{x, y\} \in S$
>   implies
> $f(\{x, y\}) = (x, y)$ or $f(\{x, y\}) = (y, x)$.

Observe that from a given non-empty set of branches S, exactly 2^{pS} incidence functions of S can be manufactured.

If f is an incidence function of S and if L is a non-empty subset of branches of S, then $f|L$ is obviously an incidence function of L. We cite this trivial fact as the following Theorem, but omit the proof.

Theorem. *Let f be an incidence function of S. Let $L \subset S$ such that $L \neq 0$.*
Then,
  $f|L$ is an incidence function of L.

Finally, observe that if f is an incidence function of S, we must have $S = \operatorname{dmn} f$, and thus, the S is superfluous. This leads to the formal definition of an incidence function.

f is an incidence function

if and only of

f is an incidence function of $\operatorname{dmn} f$.

3.2 Matrices and Arrays

The reader may be distressed to learn that in this book we shall distinguish carefully between a matrix and the array of a matrix, although elsewhere in the literature such a distinction is usually avoided and perhaps even considered pedantic. Usually, in the literature, if $m \in \omega$ and $n \in \omega$, the rectangular array of the mn objects, M_{ij}, for $i \in [m]$ and $j \in [n]$,

$$\begin{pmatrix} M_{11} & M_{12} & \dots & M_{1n} \\ M_{21} & M_{22} & \dots & M_{2n} \\ \vdots & & & \\ M_{m1} & M_{m2} & \dots & M_{mn} \end{pmatrix}$$

with appropriately decorative brackets surrounding the array is called an m by n matrix M. However, the thing of significance in the above array is not the position of each object in the array, but merely the rule specifying each object M_{ij} for $i \in [m]$ and $j \in [n]$. Once we know the mn objects M_{ij} for $i \in [m]$ and $j \in [n]$, the array is just a convenient way to display this information.

But the statement that the mn objects, M_{ij}, for $i \in [m]$ and $j \in [n]$, are known is equivalent to the fact that M is actually a function with domain $([m] \times [n])$ subject to the notational agreement that we write M_{ij} instead of $M(i, j)$ whenever $i \in [m]$ and $j \in [n]$. We shall define an m by n matrix as such a function. Formally,

M is an m by n matrix
if and only if

(1) $m \in \omega$ and $n \in \omega$.
(2) M is a function.
(3)   dmn $M = ([m] \times [n])$.

If M is an m by n matrix we shall frequently find it convenient to write M_{ij} instead of $M(i, j)$ if $i \in [m]$ and $j \in [n]$, but many times we shall prefer to use the customary functional notation $M(i, j)$.

If M is an m by n matrix we shall occasionally display the array of the mn objects, M_{ij}, for $i \in [m]$ and $j \in [n]$, but we shall eliminate the decorative brackets. In such a display the array will consist of m rows and n columns, and this array will be called the array of the matrix M.

By a matrix we mean a function M such that for some m and for some n, M is an m by n matrix. Formally,

M is a matrix
if and only if
for some m and for some n
M is an m by n matrix.

If M is a matrix we can describe m and n such that M is an m by n matrix in terms of the domain of M. It is easy to see that m is equal to the number of elements of dmn dmn M while n is equal to the number of elements of rng dmn M. We cite this fact as a Theorem, but omit the proof.

Theorem. *Let M be a matrix.*
Then,
   M is a $p$dmn dmn M by $p$rng dmn M matrix.

3.3 Submatrices

Because of the definition of a matrix adopted in 3.2, some complications enter into our definition of a submatrix. Our situation is analogous to the problem arising in the proper definition of a subsequence. Most readers are probably aware that if f is a finite sequence, then a subsequence of f is a composition function $f \circ h$, where h is a finite increasing sequence with rng $h \subset$ dmn f.
Thus, for each $i \in$ dmn h,

$$(f \circ h)(i) = f(h(i)) = f(h_i) = f_{h_i}.$$

and the subsequence $f \circ h$ essentially picks out from rng f the objects f_{h_i}; for $i \in$ dmn h, to be retained.

In the case of a submatrix of a matrix we must introduce two subsequences, one for each argument. The subsequence introduced into the first coordinate picks out the rows of the array of the original matrix to be retained in the array of the submatrix, and the subsequence introduced into the second coordinate picks out the columns of the array of the original matrix to be retained in the array of the submatrix.

We proceed to make such considerations precise. To begin, we define an increasing sequence.

> f is an increasing sequence
>
> if and only if
>
> (1) f is a finite sequence.
>
> (2) rng $f \subset R$.
>
> (3) $i \in$ dmn f and $j \in$ dmn f and $i < j$
> implies
> $f_i < f_j$.

Let $m \in \omega$. It will be convenient for us to have a symbol for the set of all those increasing sequences f such that rng f is a subset of $[m]$. We adopt the symbol $\mathscr{I}_m$ for such a set of sequences. Formally,

$$\mathscr{I}_m = \{f \mid f \text{ is an increasing sequence and } \operatorname{rng} f \subset [m]\}.$$

Let M be an m by n matrix. Let $u \in \mathscr{I}_m$ and let $v \in \mathscr{I}_n$. We want to manufacture from M a new matrix such that the array of the new matrix will consist of those rows of the array of M numbered u_i, for $i \in [pu]$, and those columns of M numbered v_j, for $j \in [pv]$. This new manufactured matrix will be denoted by M_v^u, and we call such a matrix M_v^u a submatrix of M. Thus, M_v^u is a pu by pv matrix such that

$$M_v^u(i, j) = M(u_i, v_j),$$

for each $i \in [pu]$ and each $j \in [pv]$. Formally, M_v^u can be defined as the following set of ordered pairs.

$$M_v^u = \{((i, j), M(u_i, v_j)) \mid i \in [pu] \text{ and } j \in [pv]\}.$$

We can now define a submatrix precisely by stating that K is a submatrix of M if and only if there exist appropriate u and v such that $K = M_v^u$. Formally,

K is a submatrix of M

if and only if

M is a matrix, and

for some u and for some v,

(1) $u \in \mathscr{I}_{p\,\mathrm{dmn\,dmn}\,M}$.

(2) $v \in \mathscr{I}_{p\,\mathrm{rng\,dmn}\,M}$.

(3) $M_v^u = K$.

Again let M be an m by n matrix and let $u \in \mathscr{I}_m$. Noting 1.8, (5),

$$M^u_{I_{[n]}}$$

is that submatrix of M whose array consists of those rows of the array of M numbered u_i, for $i \in [pu]$, and all columns of the array of M. To simplify the notation we shall denote such a submatrix of M by M^u. Formally,

$$M^u = M^u_{I_{[p\,\mathrm{rng\,dmn}\,M]}}.$$

Similarly, we let

$$M_v = M^{I_{[p\,\mathrm{dmn\,dmn}\,M]}}_v.$$

Fortunately, in spite of the obstacle furnished by our elaborate definition, the property of being a submatrix is transitive. More precisely, if K is a submatrix of M and if N is a submatrix of K, then N is a submatrix of M. We cite this fact as the following Theorem, but omit the proof, since only mechanical applications of the definitions are involved.

Theorem. *Let K be a submatrix of M. Let N be a submatrix of K. Then,*

N is a submatrix of M.

Let $m \in \omega$ and let $n \in \omega$ and let M be an $m + 1$ by $n + 1$ matrix. Let $j \in [m + 1]$ and let $k \in [n + 1]$. A submatrix of M which is of frequent interest is that submatrix of M whose array is obtained from the array of M by deleting the jth row and k^{th} column from the array of M. In the literature, the determinant of this submatrix is usually called the jk^{th} minor of M. However, we shall depart from tradition and call the submatrix itself the jk^{th} minor of M. Thus, in our terminology, the jk^{th} minor of M is the submatrix

$$M^{\{(i,\,i)\,|\,i \in [m] \text{ and } i < j\} \cup \{(i,\,i+1)\,|\,i \in [m] \text{ and } i \geq j\}}_{\{(i,\,i)\,|\,i \in [n] \text{ and } i < k\} \cup \{(i,\,i+1)\,|\,i \in [n] \text{ and } i \geq k\}}.$$

Examine the superscript in the above representation of the submatrix. The superscript is that increasing sequence $u \in \mathscr{I}_{m+1}$ such that $\mathrm{dmn}\,u = [m]$ and $u(i) = i$, if $i \in [m]$ and $i < j$, while $u(i) = i + 1$, if $i \in [m]$ and $i \geq j$.

It is convenient for us to have an abbreviation for an increasing sequence of this type, and thus, we formally define, for any $m \in \omega$ and any $j \in \omega$,

$$(m * j) = \{(i, i) | i \in [m] \text{ and } i < j\} \cup \{(i, i + 1) | i \in [m] \text{ and } i \geq j\}.$$

With this notational abbreviation introduced, the jk^{th} minor of the $m + 1$ by $n + 1$ matrix M is more conveniently written as

$$M_{(n * k)}^{(m * j)}.$$

3.4 Determinants

By a real matrix we mean a matrix whose range consists of real numbers. Formally,

M is a real matrix

if and only if

(1) M is a matrix.

(2)   rng $M \subset R$.

By a square matrix we mean a real matrix whose array consists of the same number of rows and columns. Formally,

M is a square matrix

if and only if

(1) M is a real matrix.

(2)   dmn dmn M = rng dmn M.

For each square matrix M a number called the determinant of M can be manufactured by several well-known processes. We assume that the reader is familiar with at least the elementary manufacturing processes which yield the determinant of M, and we agree to write

$$\det(M)$$

as an abbreviation for the determinant of M.

Let $m \in \omega$ and let M be an $m + 1$ by $m + 1$ real matrix. Among those elementary manufacturing processes yielding $\det(M)$ with which we assume the reader is familiar are various expansions in terms of the determinants of certain minors of M. It is worthwhile to recall two such expansions in terms of the notations introduced in 3.3. In particular, let $k \in [m + 1]$. Then,

$$\det(M) = \sum_{j \in [m+1]} (-1)^{j+k} M_{kj} \det\left(M_{(m \ast j)}^{(m \ast k)}\right),$$

and

$$\det(M) = \sum_{j \in [m+1]} (-1)^{j+k} M_{jk} \det\left(M_{(m \ast k)}^{(m \ast j)}\right).$$

3.5 Incidence Matrices

Let f be an incidence function of a set of branches S. From f and S we want to manufacture a $p\sigma S$ by pS matrix M to be called an incidence matrix. However, before this can be accomplished we must order the $p\sigma S$ vertices of σS and the pS branches of S.

Now, an ordering of any finite, non-empty set A is merely a function u such that $\operatorname{dmn} u = [pA]$ and $\operatorname{rng} u = A$. Formally,

> u is an ordering of A
> if and only if
>
> (1) A is finite and $A \neq 0$.
> (2) u is a function.
> (3) $\operatorname{dmn} u = [pA]$.
> (4) $\operatorname{rng} u = A$.

With this established, let x be an ordering of σS and let b be an ordering of S. Let M be that real $p\sigma S$ by pS matrix such that for each $i \in [p\sigma S]$ and each $j \in [pS]$, $M_{ij} = 0$ if the vertex x_i is not an element of the branch b_j, $M_{ij} = 1$ if the vertex x_i is the first coordinate of the ordered pair $f(b_j)$, and $M_{ij} = -1$ if the vertex x_i is the second coordinate of the ordered pair $f(b_j)$.

The construction of M appears to depend upon S, f, x, and b, but since $S = \operatorname{dmn} f$ and $p\sigma S = px$ and $pS = pb$, there is no need to incorporate the dependence upon S into the formal definition. Thus, we define formally,

> M is an incidence matrix of f with respect to (x, b) if and only if
>
> (1) f is an incidence function.
> (2) x is an ordering of $\sigma \operatorname{dmn} f$.
> (3) b is an ordering of $\operatorname{dmn} f$.
> (4) M is the px by pb matrix such that
>   for each $i \in [px]$ and each $j \in [pb]$,
>   (i) $M_{ij} = 0$ if $x_i \notin b_j$.
>   (ii) $M_{ij} = 1$ if x_i is the first coordinate of $f(b_j)$.
>   (iii) $M_{ij} = -1$ if x_i is the second coordinate of $f(b_j)$.

Suppose that M is an incidence matrix of f with respect to (x, b). Let $S = \text{dmn } f$, and observe that each branch of S has exactly two vertices of S as its elements. Thus, in the array of M, each column has exactly two non-zero entries, one of which is a 1 and the other of which is a -1. This trivial observation is of fundamental importance in our study of incidence matrices, and we cite it as the following Theorem, but omit the proof.

Theorem. *Let M be an incidence matrix of f with respect to (x, b). Let $j \in [pb]$.*
Then,

   (1) $p \{i | i \in [px] \text{ and } M_{ij} \neq 0\} = 2.$
   (2) $p \{i | i \in [px] \text{ and } M_{ij} = 1\} = 1.$
   (3) $p \{i | i \in [px] \text{ and } M_{ij} = -1\} = 1.$

Occasionally we shall find it convenient to refer to an incidence matrix without specifying the incidence function or the orderings of the branches and vertices involved. With this in mind, we define formally,

> M is an incidence matrix
>
> if and only if
>
> for some f and for some x and for some b,
>
> M is an incidence matrix of f with respect to (x, b).

3.6 Square Submatrices of an Incidence Matrix

Let M be an incidence matrix. In the following paragraphs we shall be concerned with the determinants of certain square submatrices of M. As our first effort in this direction, suppose that N is such a square submatrix of M that no column of the array of N has exactly one non-zero entry. In the following Lemma we show that in this case, $\det(N) = 0$.

Lemma. *Let M be an incidence matrix. Let M be an m by m submatrix of M. Suppose that $p \{i | i \in [m] \text{ and } N_{ij} \neq 0\} \neq 1$ for each $j \in [m]$. Then,*
$\det(N) = 0.$

Proof: If, for some $j \in [m]$, $p \{i | i \in [m] \text{ and } N_{ij} \neq 0\} = 0$, then $N_{ij} = 0$ for each $i \in [m]$, which is sufficient to guarantee that $\det(N) = 0$. Thus, we assume that $p \{i | i \in [m] \text{ and } N_{ij} \neq 0\} \neq 0$ for each $j \in [m]$.

The above assumption, coupled with the hypothesis, guarantees that for each $j \in [m]$,

$$p\,\{i\,|\,i \in [m] \text{ and } N_{ij} = 1\} = 1, \text{ and}$$
$$p\,\{i\,|\,i \in [m] \text{ and } N_{ij} = -1\} = 1.$$

Thus, for each $j \in [m]$,

$$\sum_{i \in [m]} N_{ij} = 0.$$

By the elementary theory of determinants, this latter fact is sufficient to guarantee that $\det(N) = 0$.

3.7 Unimodular Matrices

Let M be an incidence matrix. We know that each element of rng M is equal to 0, 1, or -1. However, we show in the following Theorem that M has the additional property that the determinant of every square submatrix of M is equal to 0, 1, or -1. Matrices with this property are called unimodular.

Theorem. *Let M be an incidence matrix. Let N be a square submatrix of M.*
Then,

$$\det(N) = 0 \text{ or } |\det(N)| = 1.$$

Proof: We prove by induction that if K is an m by m submatrix of M, then $\det(K) = 0$ or $|\det(K)| = 1$.

Clearly, if K is a 1 by 1 submatrix of M, then $\det(K) = 0$ or $|\det(K)| = 1$.

Thus, let $m \in \omega$ and let K be an $m + 1$ by $m + 1$ submatrix of M. Also, assume that if L is an m by m submatrix of M, then $\det(L) = 0$ or $|\det(L)| = 1$.
We shall show that $\det(K) = 0$ or $|\det(K)| = 1$.

If, for each $j \in [m + 1]$, $p\,\{i\,|\,i \in [m + 1] \text{ and } K_{ij} \neq 0\} \neq 1$, the Lemma of 3.6 guarantees that $\det(K) = 0$, and we are through.

Thus, produce $j \in [m + 1]$ such that $p\,\{i\,|\,i \in [m + 1] \text{ and } K_{ij} \neq 0\} = 1$. This yields $r \in [m + 1]$ such that $|K_{rj}| = 1$, and $K_{ij} = 0$ for each $i \in [m + 1]$ with $i \neq r$.

Consider $K^{(m \bullet r)}_{(m \bullet j)}$, the rj^{th} minor of K. Use of the second formula cited in 3.4 gives

$$|\det(K)| = |K_{rj}|\,\big|\det\big(K^{(m \bullet r)}_{(m \bullet j)}\big)\big| = \big|\det\big(K^{(m \bullet r)}_{(m \bullet j)}\big)\big|.$$

But by the Theorem of 3.3, $K^{(m \bullet r)}_{(m \bullet j)}$ is an m by m submatrix of M. Thus, by the inductive hypothesis,

$$\det\big(K^{(m \bullet r)}_{(m \bullet j)}\big) = 0 \text{ or } \big|\det\big(K^{(m \bullet r)}_{(m \bullet j)}\big)\big| = 1,$$

completing the proof.

3.8 Laplacian Expansion of a Determinant

In the proof of a Theorem in the next section we shall make use of the Laplacian expansion of a determinant which is a generalization of the formula of 3.4. Before describing the Laplacian expansion of a determinant, we must introduce the notion of the complement of an increasing sequence.

Let $m \in \omega$ and let $u \in \mathscr{I}_m$ such that $pu < m$. By the complement of u in $\mathscr{I}_m$ we mean that unique sequence $v \in \mathscr{I}_m$ which selects those elements of $[m]$ not in rng u. More precisely, the complement of u in $\mathscr{I}_m$ is that unique $v \in \mathscr{I}_m$ such that

(1) dmn $v = [m - pu]$.

(2) $v_1 = \min([m] - \mathrm{rng}\, u)$.

(3) For each $i \in \mathrm{dmn}\, v$ with $1 < i$,
$$v_i = \min\left([m] - \mathrm{rng}\,(u \cup (v|[i-1]))\right).$$

We shall let

$$C_m(u)$$

denote the complement of u in $\mathscr{I}_m$.

Now, let $m \in \omega$ and let M be an m by m real matrix. Let $u \in \mathscr{I}_m$ such that $pu < m$. Then, the Laplacian expansions of $\det(M)$ are the formulas describing $\det(M)$ by

$$\det(M) = \sum_{v \in \mathscr{I}_m \cap \{x \mid px = pu\}} \det(M_v^u)\det\left(M_{C_m(v)}^{C_m(u)}\right)(-1)^{\sum_{i \in [pu]}(u_i + v_i)}$$

and

$$\det(M) = \sum_{v \in \mathscr{I}_m \cap \{x \mid px = pu\}} \det(M_u^v)\det\left(M_{C_m(u)}^{C_m(v)}\right)(-1)^{\sum_{i \in [pu]}(u_i + v_i)}.$$

Observe that in the first of the above formulas we select the rows corresponding to rng u in the array of M, from these rows manufacture a square submatrix of M by selecting appropriate columns of the array of M corresponding to the range of any $v \in \mathscr{I}_m$ with $pu = pv$, take the determinant of this submatrix, multiply it by the determinant of that submatrix of M whose array consists of those rows and columns of the array of M not appearing in the submatrix manufactured previously from u and v, affix an appropriate $+1$ or -1, and then sum over all such $v \in \mathscr{I}_m$ with $pv = pu$. In the second formula the process is reversed with regard to rows and columns.

3.9 Reduced Incidence Matrix of a Tree

Let M be an incidence matrix of f with respect to (x, b) and let $T = \operatorname{dmn} f$. Suppose that T is a tree. Observe that M is a $pT + 1$ by pT matrix. From M exactly $pT + 1$ pT by pT submatrices of M can be manufactured, and the array of each such pT by pT submatrix of M is obtained from the array of M by deleting a row from the array of M. Each of these $pT + 1$ pT by pT submatrices of M is usually called a reduced incidence matrix, and the following Theorem 1 points out that the determinant of each of these $pT + 1$ pT by pT reduced incidence matrices is not equal to zero.

Theorem 1. *Let M be an incidence matrix of f with respect to (x, b). Suppose that* $\operatorname{dmn} f$ *is a tree. Let* $j \in [p \operatorname{dmn} f + 1]$. *Then,*

$$\det (M^{(p\,\mathrm{dmn}\,f\,*\,j)}) \neq 0.$$

Proof: Let $T = \operatorname{dmn} f$. Let $m = pT$. Let $K = M^{(m\,*\,j)}$. We lose no generality by assuming that $j = m + 1$, implying that $(m * j) = (m * m + 1) = I_{[m]}$.

We shall establish the existence of two finite sequences, u and v, each with domain equal to $[m]$ such that for each $i \in [m]$,

$$(1) \quad u_i \in \mathscr{I}_m \text{ and } \operatorname{dmn} u_i = [i].$$
$$(2) \quad v_i \in \mathscr{I}_m \text{ and } \operatorname{dmn} v_i = [i].$$
$$(3) \quad x_{m+1} \in \sigma\operatorname{rng} (b \circ v_i).$$
$$(4) \quad \operatorname{rng} (b \circ v_i) \text{ is connected.}$$
$$(5) \quad \det (K_{v_i}^{u_i}) \neq 0.$$

If these sequences u and v can be established, observe that $u_m = v_m = I_{[m]}$ and

$$K_{v_m}^{u_m} = K_{I_{[m]}}^{I_{[m]}} = K,$$

and thus, by (5), $\det (K) \neq 0$. Hence, to complete the proof we need only to contruct u and v inductively.

Since T is connected, $x_{m+1} \in \sigma T = \sigma\operatorname{rng} b$.
Produce $w \in [m]$ such that $x_{m+1} \in b_w$.
Also, produce $c \in [m]$ such that $b_w = \{x_c, x_{m+1}\}$.
Let $u_1 = \{(1, c)\}$ and let $v_1 = \{(1, w)\}$, and check easily that conditions (1)–(4) are satisfied.
Condition (5) is also satisfied because $x_c \in b_w$ implies

$$0 \neq M(c, w) = M(u_1(1), v_1(1))$$
$$= K_{v_1}^{u_1}(1,1) = \det (K_{v_1}^{u_1}).$$

Thus, let $n \in \omega$ such that $n + 1 \in [m]$ and assume that for each $i \in [n]$, u_i and v_i exist satisfying conditions (1)–(5). We shall construct u_{n+1} and v_{n+1} satisfying conditions (1)–(5).

Again, we lose no generality by assuming that the orderings x and b are such that $u_n = v_n = I_{[n]}$.
Observe that rng $(b \circ v_n) \subset T$, but

$$p\,\mathrm{rng}\,(b \circ v_n) = n < m = pT.$$

Thus, there exists a branch $\{y, z\}$ such that

$$\{y, z\} \in T,$$
$$\{y, z\} \notin \mathrm{rng}\,(b \circ v_n),$$
$$\{y, z\} \cap \sigma\,\mathrm{rng}\,(b \circ v_n) \neq 0,$$

for otherwise the connectedness of T would be contradicted.

Since $\{y, z\} \notin \mathrm{rng}\,(b \circ v_n)$, produce $s \in [m] - [n]$ such that $\{y, z\} = b_s$. Let

$$v_{n+1} = v_n \cup \{(n + 1, s)\},$$

and note that

$$v_{n+1} \in \mathcal{J}_m,$$
$$\mathrm{dmn}\, v_{n+1} = [n + 1],$$
$$x_{m+1} \in \sigma\,\mathrm{rng}\,(b \circ v_n) \subset \sigma\,\mathrm{rng}\,(b \circ v_{n+1}).$$

Also, since

$$\mathrm{rng}\,(b \circ v_{n+1}) = \mathrm{rng}\,(b \circ v_n) \cup \{\{y, z\}\},$$

rng $(b \circ v_{n+1})$ is connected.

Thus, we need only to construct $u_{n+1} \in \mathcal{J}_m$ such that $\mathrm{dmn}\, u_{n+1} = [n + 1]$ and $\det \left(K_{v_n+1}^{u_n+1} \right) \neq 0$.

Now, the selection of $\{y, z\}$ assures us that $y \in \sigma\,\mathrm{rng}\,(b \circ v_n)$ or $z \in \sigma\,\mathrm{rng}\,(b \circ v_n)$.
Suppose first that $y \in \sigma\,\mathrm{rng}\,(b \circ v_n)$ and $z \in \sigma\,\mathrm{rng}\,(b \circ v_n)$. Since rng $(b \circ v_n)$ is connected, produce g such that g is a proper path from y to z in rng $(b \circ v_n)$. Then, check that rng $g \cup \{\{y, z\}\}$ is a loop and rng $g \cup \{\{y, z\}\} \subset T$, which is impossible.
Thus, assume that $y \in \sigma\,\mathrm{rng}\,(b \circ v_n)$ and $z \notin \sigma\,\mathrm{rng}\,(b \circ v_n)$.

But $z \in \sigma T = \mathrm{rng}\, x$.
Produce $k \in [m + 1]$ such that $z = x_k$.
Since $z \notin \sigma\,\mathrm{rng}\,(b \circ v_n)$, note that for each $i \in [n]$,

$$x_k = z \notin (b \circ v_n)_i = (b \circ I_{[n]})_i = b_i,$$

implying that $M(k, i) = 0$, for each $i \in [n]$.

If $k \in [n]$, then for each $i \in [n]$,

$$0 = M(k, i) = M(u_n(k), v_n(i)) = K_{v_n}^{u_n}(k, i),$$

implying that $\det(K_{v_n}^{u_n}) = 0$, contradicting the inductive hypothesis with regard to u_n and v_n.

Thus, $k \notin [n]$.

Also, since $x_k = z \notin \sigma \mathrm{rng}\,(b \circ v_n)$, while $x_{m+1} \in \sigma \mathrm{rng}\,(b \circ v_n)$, we know that $k \in [m] - [n]$.

Let

$$u_{n+1} = u_n \cup \{(n + 1, k)\},$$

and note that $u_{n+1} \in \mathscr{I}_n$ and $\mathrm{dmn}\,u_{n+1} = [n + 1]$.

Also, since $x_k = z \in \{y, z\} = b_s$,

$$0 \neq M(k, s) = M(u_{n+1}(n + 1), v_{n+1}(n + 1))$$
$$= K_{v_{n+1}}^{u_{n+1}}(n + 1, n + 1).$$

But we have established that for each $i \in [n]$,

$$0 = M(k, i) = M(u_{n+1}(n + 1), v_{n+1}(i))$$
$$= K_{v_{n+1}}^{u_{n+1}}(n + 1, i).$$

Thus, by the inductive hypothesis that $\det(K_{v_n}^{u_n}) \neq 0$, and the first formula of 3.4,

$$|\det(K_{v_{n+1}}^{u_{n+1}})| = |K_{v_{n+1}}^{u_{n+1}}(n + 1, n + 1)| \, |\det((K_{v_{n+1}}^{u_{n+1}})_{(n \ast n+1)}^{(n \ast n+1)})|$$
$$= |\det(K_{v_n}^{u_n})| \neq 0.$$

The proof is complete.

Observe that in the proof of the preceding Theorem, with M, f, x, b, T, and m as specified, we made no use of the direction assigned to any branch of T by the incidence function f. Thus, the proof did not depend upon the algebraic sign of any non-zero element of $\mathrm{rng}\,M$. Indeed, if we had defined an $m + 1$ by m matrix N such that for each $i \in [m + 1]$ and each $j \in [m]$,

$$N_{ij} = 0 \text{ if } x_i \notin b_j,$$
$$N_{ij} = 1 \text{ if } x_i \in b_j,$$

then the argument of the proof shows that the determinant of any m by m submatrix of N, whose array is obtained by deleting a row from the array of N, is not equal to zero. Such a matrix N could be considered as a more general incidence matrix of T, without reference to direction, but we shall have no use for such matrices in this book.

The converse of Theorem 1 is also true. In particular, let S be a set of branches such that $p\sigma S = pS + 1$. Let M be an incidence matrix of f with respect to (x, b) and suppose that $S =$ dmn f. The converse of Theorem 1 is the propositition that if each of the $pS + 1$ pS by pS reduced incidence matrices, whose array is obtained by deleting a row from the array of M, has non-zero determinant, then, S is a tree. However, in the following Theorem 2 we prove the stronger proposition that if one of the $pS + 1$ pS by pS reduced incidence matrices has non-zero determinant, then, S is a tree.

Theorem 2. *Let M be an incidence matrix of f with respect to (x, b). Suppose that $p\sigma$ dmn $f = p$ dmn $f + 1$. Let $j \in [p$ dmn $f + 1]$. Suppose that $\det(M^{(p\,\mathrm{dmn}\,f \ast j)}) \neq 0$.*
Then,
   dmn f *is a tree.*
   Proof: Let $S =$ dmn f.
Let $m = pS$.
Let $K = M^{(m \ast j)}$.
We lose no generality by assuming that $j = m + 1$, implying that $(m \ast j) = (m \ast m + 1) = I_{[m]}$.
   By Theorem 2 of 2.10, it suffices to show that S is connected.
Suppose that S is not connected.
We complete the proof by showing that $\det(K) = 0$, which is false.
   Since S is not connected, produce F such that F is a component of S and $x_{m+1} \notin \sigma F$.
Thus, $\sigma F \subset \mathrm{rng}\,(x|[m])$.
Produce $u \in \mathcal{I}_m$ such that $\mathrm{rng}\,(x \circ u) = \sigma F$.
   But $x_{m+1} \in \sigma S$ and $x_{m+1} \notin \sigma F$.
Produce z such that $\{x_{m+1}, z\} \in S$ and $\{x_{m+1}, z\} \notin F$ Since. F is a component of S, $z \notin \sigma F = \mathrm{rng}\,(x \circ u)$, implying that $pu < m$.
   Refer now to the first formula exhibited in 3.8. To show that $\det(K) = 0$, we select any $v \in \mathcal{I}_m$ such that $pv = pu$, and show that $\det(K_v^u) = 0$.
   Let $n = pu$.
If, for each $k \in [n]$, $p\,\{i | i \in [n]$ and $(K_v^u)_{ik} \neq 0\} \neq 1$, the Lemma of 3.6 guarantees that $\det(K_v^u) = 0$, and we are through. Thus, assume that for some $k \in [n]$, $p\,\{i | i \in [n]$ and $(K_v^u)_{ik} \neq 0\} = 1$.
We shall show that this is impossible.
   Produce $r \in [n]$ such that $|(K_v^u)_{rk}| = 1$, and $(K_v^u)_{ik} = 0$ for each $i \in [n]$ with $i \neq r$.
Observe that

$$0 \neq (K_v^u)_{rk} = K(u_r, v_k) = M(u_r, v_k),$$

implying that $x_{u_r} \in b_{v_k}$.

But F is a component of S and $x_{u_r} \in \text{rng} (x \circ u) = \sigma F$. Thus, $x_{u_r} \in b_{v_k} \in F$.
Produce y such that $\{x_{u_r}, y\} = b_{v_k} \in F$, implying that $y \in \sigma F = \text{rng} (x \circ u)$.

Thus, produce $t \in [n]$ with $t \neq r$ such that $y = x_{u_t}$. But then,
$x_{u_t} \in \{x_{u_r}, x_{u_t}\} = b_{v_k}$, implying that

$$0 \neq M(u_t, v_k) = K(u_t, v_k) = (K_v^u)_{tk}.$$

This contradicts the fact that $(K_v^u)_{ik} = 0$ for each $i \in [n]$ with $i \neq r$.
The proof is complete.

3.10 Incidence Matrix of a Loop

Let M be an incidence matrix of f with respect to (x, b) and suppose
that dmn f is a loop. By (2) of the Theorem of 2.4, M is a square matrix,
and we investigate $\det(M)$. But by (1) of the Theorem of 3.5, each
column of the array of M has exactly two non-zero entries. Hence,
the Lemma of 3.6 guarantees that $\det(M) = 0$. We cite these facts as
the following Theorem, but omit the proof.

Theorem. *Let M be an incidence matrix of f with respect to (x, b).
Suppose that dmn f is a loop.
Then,*

(1)   *M is a square matrix.*
(2) $\det(M) = 0$.

CHAPTER FOUR

Linear Algebra Review

4.0 Introduction

The preceding three Chapters complete our discussion of elementary network theory. At this point the reader might legitimately expect to begin the electric circuit theory as advertised in the title of this book. However, in our formal presentation, electric circuit theory will be offered as a combination of the preceding network theory with linear algebra. Thus, before starting the electric circuit theory we must establish some preliminary material dealing with linear algebra. This Chapter is devoted to an exposition of the linear algebra needed in the remainder of the book.

Although we developed our network theory from first principles, we shall assume that the reader has some rudimentary background in linear algebra. Our requirements of the reader in this respect are quite modest. We shall demand only that minimal knowledge of linear algebra which is now required in many modern undergraduate science and engineering curricula. Indeed, the present popularity of linear algebra in the undergraduate curriculum suggests that our linear algebra requirements are no more unreasonable than the requirement of a knowledge of calculus and differential equations.

4.1 The Field of Scalars

In this Chapter we agree to let F denote some fixed field of scalars to be used on our linear algebra. Later in the book we shall specialize F as needed for various applications. For example, in our study of resistive networks we shall let F be equal to R, the field of real numbers. For work beyond the scope of this book dealing with networks with capacitive and inductive elements, F must be specialized to the field of complex numbers.

4.2 Addition and Scalar Multiplication of Functions

Let B be a linear space. Let f and g be functions such that $\operatorname{rng} f \subset B$ and $\operatorname{rng} g \subset B$. We want to add the functions f and g to manufacture a new function $f + g$ such that for each $x \in \operatorname{dmn} f \cap \operatorname{dmn} g$, $(f + g)(x)$ is equal to the element $f(x) + g(x)$ of B. Formally,

$$f + g = \{(x, f(x) + g(x)) \,|\, x \in \operatorname{dmn} f \cap \operatorname{dmn} g\}.$$

Observe that in the special case when $\operatorname{dmn} f \cap \operatorname{dmn} g$ is empty, then $f + g$ reduces to the empty set which is a perfectly respectible function whose range is a subset of B. Thus, we are assured that the manufactured function $f + g$ is always a function whose range is a subset of B.

With B and f as above, let $b \in F$. We want to multiply the function f on the left by the scalar b to manufacture a new function bf such that for each $x \in \operatorname{dmn} f$, $(bf)(x)$ is equal to the element $bf(x)$ of B. Formally,

$$bf = \{(x, bf(x)) \,|\, x \in \operatorname{dmn} f\}.$$

Such a multiplication, producing the new function bf whose range is a subset of B, is usually called the scalar multiplication of f by b.

The functional addition defined above is both commutative and associative. We cite these facts as the following Theorem 1, but omit the trivial proof.

Theorem 1. *Let B be a linear space. Let f be a function such that* $\operatorname{rng} f \subset B$. *Let g be a function such that* $\operatorname{rng} g \subset B$. *Let h be a function such that* $\operatorname{rng} h \subset B$.
Then,

(1) $f + g = g + f$.
(2) $(f + g) + h = f + (g + h)$.

When adding several functions we shall use the $\sum$ notation in the obvious way.

The scalar multiplication defined above is associative; it is also distributive with respect to functional addition. We cite these facts as the following Theorem 2, but omit the trivial proof.

Theorem 2. *Let B be a linear space. Let f be a function such that* $\operatorname{rng} f \subset B$. *Let g be a function such that* $\operatorname{rng} g \subset B$. *Let $b \in F$ and let* $c \in F$.
Then,

(1) $(bc) f = b(cf)$.
(2) $b(f + g) = bf + bg$.

Again, let B be a linear space and let f and g be functions such that $\operatorname{rng} f \subset B$ and $\operatorname{rng} g \subset B$. We want to subtract the function g from the function f in the obvious way to manufacture a new function $f - g$. Formally,

$$f - g = \{(x, f(x) - g(x)) \,|\, x \in \text{dmn } f \cap \text{dmn } g\}.$$

Finally, with B and f as above, we note that $-f$ has not yet been defined. Thus, formally, we let

$$-f = (-1)\,f.$$

4.3 Linear Space of 0-Chains

Let K be any set. Let $\mathscr{L}(K)$ be the set of all functions whose domain is equal to K and whose range is a subset of F. Observe that F itself is a linear space, and thus, by 4.2, addition and scalar multiplication has been defined for the functions of $\mathscr{L}(K)$. With such a definition of addition and scalar multiplication it is easy to verify that $\mathscr{L}(K)$ is itself a linear space which we call the space of 0-chains of K. We define formally,

$$\mathscr{L}(K) = \{f \,|\, f \text{ is a function and dmn } f = K \text{ and rng } f \subset F\}.$$

In the following Theorem we document the fact that $\mathscr{L}(K)$ is a linear space, but omit the proof.

Theorem. *Let K be any set.*
Then,
   $\mathscr{L}(K)$ is a linear space.

4.4 Canonical Base of the Space of 0-Chains

Let K be any set and let $y \in K$. We consider that 0-chain of $\mathscr{L}(K)$ which takes on the value 1 at that point $y \in K$, but takes on the value 0 at all other points of K. We denote this 0-chain by Φ_y^K. Formally,

$$\Phi_y^K = \{(y, 1)\} \cup (K - \{y\} \times \{0\}).$$

Suppose now that K is finite and non-empty. Then, the set of all such 0-chains defined above, $\{\Phi_y^K \,|\, y \in K\}$, generates the whole space of 0-chains $\mathscr{L}(K)$. By this we mean that any $g \in \mathscr{L}(K)$ can be expressed as a linear combination of the set $\{\Phi_y^K \,|\, y \in K\}$; the coefficient of Φ_y^K in such a linear combination is equal to $g(y)$ for each $y \in K$. In the following Theorem 1 we cite this fact.

Theorem 1. *Let K be finite and non-empty. Let $g \in \mathscr{L}(K)$.*
Then,

$$g = \sum_{y \in K} g(y)\,\Phi_y^K.$$

Proof: Let $x \in K$. Since $\Phi_y^K(x) = 0$ for each $y \in K$ with $y \neq x$, while $\Phi_x^K(x) = 1$,

$$\left(\sum_{y \in K} g(y)\, \Phi_y^K \right)(x) = \sum_{y \in K} g(y)\, \Phi_y^K(x) = g(x)\, \Phi_x^K(x) = g(x).$$

Since g and $\sum_{y \in K} g(y)\, \Phi_y^K$ are both functions with domain equal to K, the above suffices to complete the proof.

With K finite and non-empty, we point out in the following Theorem 2 the additional fact that our special set of 0-chains under consideration, $\{\Phi_y^K | y \in K\}$, is a base of the linear space $\mathscr{L}(K)$. This base, $\{\Phi_y^K | y \in K\}$, is usually called the canonical base of the space of 0-chains $\mathscr{L}(K)$.

Theorem 2. *Let K be finite and non-empty. Then,*

$\{\Phi_y^K | y \in K\}$ is a base of $\mathscr{L}(K)$.

Proof: Let $g \in \mathscr{L}(K)$.
Since by Theorem 1,

$$g = \sum_{y \in K} g(y)\, \Phi_y^K,$$

we need only show that the above representation is unique. Thus, suppose that h is a function with dmn $h = K$ and rng $h \subset F$ such that

$$g = \sum_{y \in K} h(y)\, \Phi_y^K.$$

We must show that $h(x) = g(x)$ for each $x \in K$.
Let $x \in K$.
Since $\Phi_y^K(x) = 0$ for each $y \in K$ with $y \neq x$, while $\Phi_x^K(x) = 1$,

$$g(x) = \left(\sum_{y \in K} h(y)\, \Phi_y^K \right)(x)$$

$$= \sum_{y \in K} h(y)\, \Phi_y^K(x) = h(x)\, \Phi_x^K(x) = h(x).$$

The proof is complete.

If B is any linear space, by

$$\dim(B)$$

we mean the dimension of B. Thus, $\dim(B)$ is either a non-negative integer or $\dim(B)$ is equal to infinity.

In particular, if K is finite and non-empty, we consider $\dim(\mathscr{L}(K))$, the dimension of the space of 0-chains. In the following Theorem 3 we point out that $\dim(\mathscr{L}(K))$ is exactly equal to the number of elements of K.

Theorem 3. *Let K be finite and non-empty.*
Then,

$$\dim(\mathscr{L}(K)) = pK.$$

Proof: By Theorem 2, $\{\Phi_y^K | y \in K\}$ is a base of $\mathscr{L}(K)$, and thus, since $\{\Phi_y^K | y \in K\}$ is finite,

$$\dim(\mathscr{L}(K)) = p\{\Phi_y^K | y \in K\}.$$

But consider the function u such that

$$u = \{(y, \Phi_y^K) | y \in K\}.$$

Clearly, u is univalent, $\operatorname{dmn} u = K$, and $\operatorname{rng} u = \{\Phi_y^K | y \in K\}$, which guarantees that $pK = p\{\Phi_y^K | y \in K\}$, completing the proof.

4.5 Inner Product

Let K be a finite, non-empty set and let $f \in \mathscr{L}(K)$ and let $g \in \mathscr{L}(K)$. We define the inner product of f and g, denoted by $\langle f, g \rangle$, as the sum for all $x \in K$ of products $f(x) g(x)$. Formally,

$$\langle f, g \rangle = \sum_{x \in K} f(x) g(x).$$

This inner product, defined above, satisfies the usual conditions implied by the name "inner product". We exhibit these conditions in the following Theorem 1, but omit the proof.

Theorem 1. *Let K be finite and non-empty. Let $f \in \mathscr{L}(K)$ and let $g \in \mathscr{L}(K)$ and let $h \in \mathscr{L}(K)$. Let $a \in F$ and let $b \in F$.*
Then,

   (1) $\langle f, g \rangle = \langle g, f \rangle$.
   (2) $\langle a f + b g, h \rangle = a \langle f, h \rangle + b \langle g, h \rangle$.

Let K be finite and non-empty and let $g \in \mathscr{L}(K)$ and let $y \in K$. Then $g(y)$ is equal to the inner product of the canonical base element Φ_y^K with g. We cite this useful fact as the following Theorem 2, but omit the trivial proof.

Theorem 2. *Let K be finite and non-empty. Let $g \in \mathscr{L}(K)$. Let $y \in K$.*
Then,

$$g(y) = \langle \Phi_y^K, g \rangle.$$

With K finite and non-empty, the canonical base $\{\Phi_y^K | y \in K\}$ of $\mathscr{L}(K)$ is an orthonormal base with respect to this inner product. We state this fact as the following Theorem 3, exhibiting explicitly the conditions of orthonormality.

Theorem 3. *Let K be finite and non-empty. Let $y \in K$ and let $x \in K$ such that $y \neq x$.*
Then,

    (1) $\langle \Phi_y^K, \Phi_x^K \rangle = 0.$

    (2) $\langle \Phi_y^K, \Phi_y^K \rangle = 1.$

Proof: To prove (1), note that by Theorem 2,

$$0 = \Phi_x^K(y) = \langle \Phi_y^K, \Phi_x^K \rangle.$$

To prove (2), note that by Theorem 2,

$$1 = \Phi_y^K(y) = \langle \Phi_y^K, \Phi_y^K \rangle.$$

In the preceding Theorem 3 we have exhibited the fact that the canonical base of $\mathscr{L}(K)$ is orthonormal with respect to our inner product when K is finite and non-empty. An important, well-known theorem in linear algebra guarantees that if M is any subspace of $\mathscr{L}(K)$ of positive dimension, then there exists an orthonormal base of M with respect to this inner product. We shall need this fact in our later work, and we exhibit it as the following Theorem 4, omitting the proof.

Theorem 4. *Let K be finite and non-empty. Let M be a subspace of $\mathscr{L}(K)$. Suppose that* $\dim(M) > 0$.
Then,

    for some B,

    (1) *B is a base of M.*

    (2) *$x \in B$ implies $\langle x, x \rangle = 1$.*

    (3) *$x \in B$ and $y \in B$ and $x \neq y$*
        implies
        $\langle x, y \rangle = 0$.

Let K be finite and non-empty and let $g \in \mathscr{L}(K)$. We can write g as a linear combination of the canonical base $\{\Phi_y^K | y \in K\}$. In particular, if $y \in K$ we have seen in Theorem 1 of 4.4 that the coefficient of Φ_y^K in this representation of g is equal to $g(y)$, and in Theorem 2 of this paragraph we have stated that this coefficient is also equal to $\langle \Phi_y^K, g \rangle$, the inner product of g with the base element Φ_y^K.

This inner product formula is valid in the case of any orthonormal base of a subspace of $\mathscr{L}(K)$ of positive dimension. In particular, suppose that M is a subspace of $\mathscr{L}(K)$ of positive dimension and let B be an orthonormal base of M. Suppose that $g \in M$ and let $b \in B$. In the following Theorem 5 we point out that the coefficient of b when g is written as a linear combination of the base B is equal to the inner product $\langle b, g \rangle$. We omit the trivial proof.

Theorem 5. *Let K be finite and non-empty. Let M be a subspace of $\mathscr{L}(K)$. Suppose that $\dim(M) > 0$. Let B be a base of M. Suppose that*

(1) $x \in B$ *implies* $\langle x, x \rangle = 1$.

(2) $x \in B$ *and* $y \in B$ *and* $x \neq y$

implies

$\langle x, y \rangle = 0$.

Let $g \in M$.
Then,

$$g = \sum_{b \in B} \langle b, g \rangle\, b.$$

4.6 Linear Maps

We shall frequently consider linear maps of one linear space into another linear space. As a reminder to the reader we define formally,

f is a linear map of B into D
if and only of

(1) f is a function.

(2) $\operatorname{dmn} f = B$ and $\operatorname{rng} f \subset D$.

(3) B is a linear space and D is a linear space.

(4) $x \in B$ and $y \in B$ implies $f(x + y) = f(x) + f(y)$.

(5) $x \in B$ and $a \in F$ implies $f(ax) = af(x)$.

We assume that the reader is familiar with linear maps and their elementary properties. It is worthwhile, however, to point out explicitly that a linear map is completely determined by its values on a finite, non-empty base. More precisely, if K is a finite, non-empty base of a linear space B and if D is a linear space, and if g is a function whose domain is equal to K and whose range is a subset of D, then g can be extended uniquely to a linear map f of B into D. This fact is exhibited as the following Theorem.

Theorem. *Let B be a linear space and let D be a linear space. Let K be a base for B such that K is finite and non-empty. Let g be a function such that $\operatorname{dmn} g = K$ and $\operatorname{rng} g \subset D$.*
Then,

there exists one and only one function f such that

(1) $g \subset f$.

(2) f *is a linear map of B into D.*

Proof: For each $x \in B$ let u_x be that function with domain K such that

$$x = \sum_{b \in K} u_x(b)\, b.$$

The existence of u_x for each $x \in B$ is guaranteed by the fact that K is a base for B.

Let f be that function with domain B such that for each $x \in B$,

$$f(x) = \sum_{b \in K} u_x(b)\, g(b).$$

It is easy to check that f satisfies the conditions required by this Theorem.

4.7 Transpose of a Linear Map

In the literature the word "transpose" is usually applied to matrices to manufacture a new matrix called the transpose of the given matrix. However, in this book we shall depart from tradition and apply the word "transpose" to linear maps, manufacturing a new linear map called the transpose of the given linear map. Actually, we shall only consider the transpose of a limited class of linear maps such that the domains of these linear maps are finite dimensional spaces of 0-chains and the ranges of these linear maps are subsets of their domains.

To be more precise let K be a finite non-empty set and let g be a linear map of $\mathscr{L}(K)$ into $\mathscr{L}(K)$. Observe that for each $y \in K$, Φ_y^K is an element of the base $\{\Phi_x^K \mid x \in K\}$ for $\mathscr{L}(K)$, and thus, by the Theorem of 4.6 there exists a unique linear map f of $\mathscr{L}(K)$ into $\mathscr{L}(K)$ such that for each $y \in K$,

$$f(\Phi_y^K) = \sum_{x \in K} \langle g(\Phi_x^K),\, \Phi_y^K \rangle\, \Phi_x^K.$$

We shall call this unique linear map f of $\mathscr{L}(K)$ into $\mathscr{L}(K)$ the transpose of g, with respect to K. Formally,

> f is the transpose of g, with respect to K
> if and only if
>
> (1) K is finite and non-empty.
>
> (2) g is a linear map of $\mathscr{L}(K)$ into $\mathscr{L}(K)$.
>
> (3) f is that unique linear map of $\mathscr{L}(K)$ into $\mathscr{L}(K)$
>   such that for each $y \in K$,

$$f(\Phi_y^K) = \sum_{x \in K} \langle g(\Phi_x^K),\, \Phi_y^K \rangle\, \Phi_x^K.$$

Suppose now that f is the transpose of g, with respect to K. We want to establish a notational abbreviation for f. Note, however, that in this situation K is completely determined by g. In fact, it is easy to show that

$$K = \mathrm{dmn}\ \sigma\ \mathrm{dmn}\ g,$$

and thus, there is no need for the appearance of K in our notational abbreviation for f. With this in mind we adopt the notation

$$g^t$$

for such a function f, where f is the transpose of g, with respect to K.

Let K be finite and non-empty and let f be a linear map of $\mathscr{L}(K)$ into $\mathscr{L}(K)$ and let g be a linear map of $\mathscr{L}(K)$ into $\mathscr{L}(K)$. If, for each $x \in K$ and each $y \in K$, $\langle g(\Phi_x^K), \Phi_y^K \rangle = \langle \Phi_x^K, f(\Phi_y^K) \rangle$, then f is the transpose of g with respect to K.

On the other hand, if f is the transpose of g with respect to K, then for each $s \in \mathscr{L}(K)$ and each $t \in \mathscr{L}(K)$, $\langle f(s), t \rangle = \langle s, g(t) \rangle$.

We exhibit these two facts as the following two Theorems.

Theorem 1. *Let K be finite and non-empty. Let f be a linear map of $\mathscr{L}(K)$ into $\mathscr{L}(K)$. Let g be a linear map of $\mathscr{L}(K)$ into $\mathscr{L}(K)$. Suppose that $x \in K$ and $y \in K$ implies $\langle g(\Phi_x^K), \Phi_y^K \rangle = \langle \Phi_x^K, f(\Phi_y^K) \rangle$.*
Then,

$$f = g^t.$$

Proof: Let $y \in K$. It suffices to show that for each $x \in K$,

$$(f(\Phi_y^K))(x) = \langle \sum_{z \in K} \langle g(\Phi_z^K), \Phi_y^K \rangle \, \Phi_z^K \rangle (x)$$

$$= \sum_{z \in K} \langle g(\Phi_z^K), \Phi_y^K \rangle \, \Phi_z^K(x) = \langle g(\Phi_x^K), \Phi_y^K \rangle.$$

Thus, let $x \in K$.
But $f(\Phi_y^K) \in \mathscr{L}(K)$.
Hence, by Theorem 2 of 4.5, and the hypothesis,

$$(f(\Phi_y^K))(x) = \langle \Phi_x^K, f(\Phi_y^K) \rangle = \langle g(\Phi_x^K), \Phi_y^K \rangle.$$

The proof is complete.

Theorem 2. *Let K be finite and non-empty. Let g be a linear map of $\mathscr{L}(K)$ into $\mathscr{L}(K)$. Let $s \in \mathscr{L}(K)$ and let $t \in \mathscr{L}(K)$.*
Then,

$$\langle g^t(s), t \rangle = \langle s, g(t) \rangle.$$

Proof: For each $r \in \mathscr{L}(K)$ let u_r be that unique fuction with domain K such that for each $r \in \mathscr{L}(K)$,

$$r = \sum_{x \in K} u_r(x) \, \Phi_x^K.$$

Then, by the linearity of g^t and properties of the inner product,

$$\langle g^t(s), t \rangle = \langle g^t \big(\sum_{x \in K} u_s(x) \, \Phi_x^K \big), \ \sum_{y \in K} u_t(y) \, \Phi_y^K \rangle$$

$$= \langle \sum_{x \in K} g^t(u_s(x) \, \Phi_x^K), \ \sum_{y \in K} u_t(y) \, \Phi_y^K \rangle$$

$$= \langle \sum_{x \in K} u_s(x) \, g^t(\Phi_x^K), \ \sum_{y \in K} u_t(y) \, \Phi_y^K \rangle$$

$$= \sum_{x \in K} \sum_{y \in K} u_s(x) \, u_t(y) \, \langle g^t(\Phi_x^K), \Phi_y^K \rangle.$$

But consider any $x \in K$ and $y \in K$.

By the definition of the transpose and properties of the inner product,

$$\langle g^t(\Phi_x^K), \Phi_y^K \rangle = \langle \sum_{z \in K} \langle g(\Phi_z^K), \Phi_x^K \rangle \Phi_z^K, \Phi_y^K \rangle$$

$$= \sum_{z \in K} \langle g(\Phi_z^K), \Phi_x^K \rangle \langle \Phi_z^K, \Phi_y^K \rangle$$

$$= \langle g(\Phi_y^K), \Phi_x^K \rangle = \langle \Phi_x^K, g(\Phi_y^K) \rangle.$$

Thus, using this fact in the preceding equation plus properties of the inner product and the linearity of g,

$$\langle g^t(s), t \rangle = \sum_{x \in K} \sum_{y \in K} u_s(x) \, u_t(y) \, \langle g^t(\Phi_x^K), \Phi_y^K \rangle$$

$$= \sum_{x \in K} \sum_{y \in K} u_s(x) \, u_t(y) \, \langle \Phi_x^K, g(\Phi_y^K) \rangle$$

$$= \langle \sum_{x \in K} u_s(x) \, \Phi_x^K, \ \sum_{y \in K} u_t(y) \, g(\Phi_y^K) \rangle$$

$$= \langle \sum_{x \in K} u_s(x) \, \Phi_x^K, \ \sum_{y \in K} g(u_t(y) \, \Phi_y^K) \rangle$$

$$= \langle \sum_{x \in K} u_s(x) \, \Phi_x^K, \ g \big(\sum_{y \in K} u_t(y) \, \Phi_y^K \big) \rangle$$

$$= \langle s, g(t) \rangle.$$

The proof is complete.

Let K be finite and non-empty and let f be a linear map of $\mathscr{L}(K)$ into $\mathscr{L}(K)$ and let g be a linear map of $\mathscr{L}(K)$ into $\mathscr{L}(K)$. Suppose that for each $s \in \mathscr{L}(K)$ and each $t \in \mathscr{L}(K)$

$$\langle g(s), t \rangle = \langle s, f(t) \rangle.$$

Then, in particular, for each $x \in K$ and each $y \in K$,

$$\langle g(\Phi_x^K), \Phi_y^K \rangle = \langle \Phi_x^K, f(\Phi_y^K) \rangle,$$

so that Theorem 1 guarantees that f is equal to the transpose of g.

This fact, combined with Theorem 2, lets us write a condition involving each $s \in \mathscr{L}(K)$ and each $t \in \mathscr{L}(K)$ which is both necessary and sufficient for f to be equal to the transpose of g. This is done in the following Theorem, but we omit the formal proof.

Theorem 3. *Let K be finite and non-empty. Let f be a linear map of $\mathscr{L}(K)$ into $\mathscr{L}(K)$. Let g be a linear map of $\mathscr{L}(K)$ into $\mathscr{L}(K)$.*
Then,

$$f = g^t$$

if and only if

$$s \in \mathscr{L}(K) \text{ and } t \in \mathscr{L}(K) \text{ implies } \langle g(s), t \rangle = \langle s, f(t) \rangle.$$

Let K be finite and non-empty and let g be a linear map of $\mathscr{L}(K)$ into $\mathscr{L}(K)$. Then, the transpose of the transpose of g is, as expected, equal to g. We cite this fact as the following Theorem, adopting the notation of g^{tt} for the transpose of the transpose of g.

Theorem 4. *Let K be finite and non-empty. Let g be a linear map of $\mathscr{L}(K)$ into $\mathscr{L}(K)$.*
Then,

$$g^{tt} = g.$$

Proof: Let $s \in \mathscr{L}(K)$ and let $t \in \mathscr{L}(K)$.
By Theorem 3, it suffices to show that

$$\langle g^t(s), t \rangle = \langle s, g(t) \rangle,$$

which is guaranteed by Theorem 2.

Again, let K be finite and non-empty and let g be a linear map of $\mathscr{L}(K)$ into $\mathscr{L}(K)$ and let $b \in F$. Then, the transpose of the linear map bg is equal to b multiplied by the transpose of g. We cite this fact as the following Theorem.

Theorem 5. *Let K be finite and non-empty. Let g be a linear map of $\mathscr{L}(K)$ into $\mathscr{L}(K)$. Let $b \in F$.*
Then,

$$(bg)^t = bg^t.$$

Proof: Let $s \in \mathscr{L}(K)$ and let $t \in \mathscr{L}(K)$.
By Theorem 3, it suffices to show that

$$\langle (bg)(s), t \rangle = \langle s, (bg^t)(t) \rangle.$$

But, by a property of the inner product and Theorem 2,

$$\begin{aligned}
\langle (bg)\,(s),\, t\rangle &= \langle bg(s),\, t\rangle \\
&= b\,\langle g(s),\, t\rangle \\
&= b\,\langle s,\, g^t(t)\rangle \\
&= \langle s,\, bg^t(t)\rangle \\
&= \langle s,\, (bg^t)\,(t)\rangle.
\end{aligned}$$

Let K be finite and non-empty and let g be a linear map of $\mathscr{L}(K)$ into $\mathscr{L}(K)$ and let h be a linear map of $\mathscr{L}(K)$ into $\mathscr{L}(K)$. Then, the transpose of the linear map $g + h$ is equal to the transpose of g added to the transpose of h. We cite this fact as the following Theorem.

Theorem 6. *Let K be finite and non-empty. Let g be a linear map of $\mathscr{L}(K)$ into $\mathscr{L}(K)$. Let h be a linear map of $\mathscr{L}(K)$ into $\mathscr{L}(K)$.*
Then,

$$(g + h)^t = g^t + h^t.$$

Proof: Let $s \in \mathscr{L}(K)$ and let $t \in \mathscr{L}(K)$.
By Theorem 3 it suffices to show that

$$\langle (g + h)\,(s),\, t\rangle = \langle s,\, (g^t + h^t)\,(t)\rangle.$$

But by a property of the inner product and Theorem 2,

$$\begin{aligned}
\langle (g + h)\,(s),\, t\rangle &= \langle g(s) + h(s),\, t\rangle \\
&= \langle g(s),\, t\rangle + \langle h(s),\, t\rangle \\
&= \langle s,\, g^t(t)\rangle + \langle s,\, h^t(t)\rangle \\
&= \langle s,\, g^t(t) + h^t(t)\rangle \\
&= \langle s,\, (g^t + h^t)\,(t)\rangle.
\end{aligned}$$

Again, let K be finite and non-empty and let g be a linear map of $\mathscr{L}(K)$ into $\mathscr{L}(K)$ and let h be a linear map of $\mathscr{L}(K)$ into $\mathscr{L}(K)$. As the next Theorem we show that the transpose of the linear map $g \circ h$ is equal to the functional composition of the transpose of h with the transpose of g, noting that the order is reversed in the functional composition of the transposes.

Theorem 7. *Let K be finite and non-empty. Let g be a linear map of $\mathscr{L}(K)$ into $\mathscr{L}(K)$. Let h be a linear map of $\mathscr{L}(K)$ into $\mathscr{L}(K)$.*
Then,

$$(g \circ h)^t = h^t \circ g^t.$$

Proof: Let $s \in \mathscr{L}(K)$ and let $t \in \mathscr{L}(K)$.
By Theorem 3, it suffices to show that

$$\langle (g \circ h)\,(s),\, t\rangle = \langle s,\, (h^t \circ g^t)\,(t)\rangle.$$

But by Theorem 2, twice,

$$\langle (g \circ h)\,(s),\,t \rangle = \langle g\,(h\,(s)),\,t \rangle$$
$$= \langle h\,(s),\,g^t\,(t) \rangle$$
$$= \langle s,\,h^t\,(g^t\,(t)) \rangle$$
$$= \langle s,\,(h^t \circ g^t)\,(t) \rangle.$$

Again, let K be finite and non-empty. Certain linear maps of $\mathscr{L}(K)$ into $\mathscr{L}(K)$ are such that each is equal to its own transpose. Such linear maps of $\mathscr{L}(K)$ into $\mathscr{L}(K)$ are called symmetric linear maps, and these symmetric linear maps will be of interest to us later in this book.

As our first example of a symmetric linear map we consider a linear map of $\mathscr{L}(K)$ into $\mathscr{L}(K)$ which, for each element x of K, maps the canonical base element Φ_x^K into a scalar multiple of itself. Such a linear map is called a diagonal linear map, and in the following Theorem 8 we show that a diagonal linear map is equal to its own transpose, and thus, symmetric.

Theorem 8. *Let K be finite and non-empty. Let u be a function such that* $\mathrm{dmn}\ u = K$ *and* $\mathrm{rng}\ u \subset R$. *Let g be that unique linear map of $\mathscr{L}(K)$ into $\mathscr{L}(K)$ such that for each $x \in K$, $g(\Phi_x^K) = u(x)\,\Phi_x^K$.*
Then,
$$g = g^t.$$
Proof: We apply Theorem 1.
Let $x \in K$ and let $y \in K$.
It suffices to show that $\langle g(\Phi_x^K),\,\Phi_y^K \rangle = \langle g(\Phi_y^K),\,\Phi_x^K \rangle$.
Since the above desired result is true when $x = y$, we may assume that $x \neq y$.
But then,

$$\langle g(\Phi_x^K),\,\Phi_y^K \rangle = \langle u(x)\,\Phi_x^K,\,\Phi_y^K \rangle = u(x)\,\langle \Phi_x^K,\,\Phi_y^K \rangle = 0.$$
$$\langle g(\Phi_y^K),\,\Phi_x^K \rangle = \langle u(y)\,\Phi_y^K,\,\Phi_x^K \rangle = u(y)\,\langle \Phi_y^K,\,\Phi_x^K \rangle = 0.$$

The proof is complete.

4.8 Direct Sum Decomposition

Let B be a linear space and let S be a subset of B and let T be a subset of B. By the sum of the subsets, $S + T$, we mean the set of all elements $x + y$ such that x is an element of S and y is an element of T. Formally,

$$S + T = \{x + y \mid x \in S \text{ and } y \in T\}.$$

In the special case when $S + T$ is equal to B, note that each element b of B can be decomposed into the sum of two elements $x + y$ such that x is an element of S and y is an element of T, but this decomposition of b is not necessarily unique.

If the decomposition of each element b of B into such a sum $x + y$ is unique, with x an element of S and y an element of T, and if, in addition, S is a subspace of B and T is a subspace of B, we say that B is decomposed into the direct sum of S and T, and we write

$$B = S \oplus T.$$

Formally,

$B = S \oplus T$
if and only if
(1) B is a linear space.
(2) S is a subspace of B.
(3) T is a subspace of B.
(4)   If $b \in B$, then there exists a
      unique (x, y) such that
      $x \in S$ and $y \in T$ and
      $b = x + y$.

Now, let B be a linear space with zero element θ. Let S be a subspace of B and let T be a subspace of B such that

$$B = S + T.$$

We want to cite a condition on S and T which will be necessary and sufficient for B to be decomposed into the direct sum of S and T. In the following Theorem we show that such a condition is the requirement that the intersection of S and T consists of only the element θ.

Theorem. *Let B be a linear space with zero element θ. Let S be a subspace of B. Let T be a subspace of B. Suppose that $B = S + T$. Then,*
   $B = S \oplus T$ if and only if $S \cap T = \{\theta\}$.

Proof: Suppose that $B = S \oplus T$.
Since we know trivially that $\{\theta\} \subset S \cap T$, it suffices to show that $S \cap T \subset \{\theta\}$.
Thus, let $b \in S \cap T$.
We shall show that $b = \theta$.

But since $b \in S$ and $b \in T$, note that we have two decompositions

$$b = b + \theta, \text{ and}$$
$$b = \theta + b,$$

of the element b as the sum of two elements, the first of which is an element of S and the second of which is an element of T. Since $B = S \oplus T$, the two decompositions must be the same, and hence, $b = \theta$.

In the other direction, suppose now that $S \cap T = \{\theta\}$. To show that $B = S \oplus T$, we suppose that there exists $u \in B$ admitting two decompositions,

$$u = x + y, \text{ and}$$

$$u = p + q,$$

with $x \in S$ and $y \in T$ and $p \in S$ and $q \in T$. It suffices to show that $x = p$ and $y = q$.

But we know that

$$x + y = u = p + q,$$

from which

$$x - p = q - y.$$

But $x \in S$ and $p \in S$, implying that $x - p \in S$.
Similarly, $q - y \in T$.
Thus, $x - p \in S \cap T = \{\theta\}$, implying that $x - p = \theta$, and $x = p$.
Similarly, $y = q$.

4.9 Dimension and Direct Sum Decomposition

Let L be a finite dimensional linear space. Let S and T be subspaces of L such that L is decomposed into the direct sum of S and T. A well-known fact of linear algebra states that the dimension of L is equal to the sum of the dimension of S and the dimension of T. We document this fact as the following Theorem, but omit the proof.

Theorem. *Let L be a finite dimensional linear space. Let S and T be such that $L = S \oplus T$.*
Then,
$$\dim(L) = \dim(S) + \dim(T).$$

CHAPTER FIVE

Boundary Operator and Coboundary Operator

5.0 Introduction

We now start to develop our circuit theory by combining our linear algebra and network theory. In this Chapter such a combination yields the boundary operator and coboundary operator, leading to a precise formulation in Chapter Six of Kirchhoff's Laws, upon which all circuit theory is based.

5.1 Assumptions of This Chapter

As in the preceding Chapter we let F be some fixed field of scalars to be used in our linear algebra.

We also fix some incidence function f.

Recall from Chapter Three that $(\sigma\,\mathrm{dmn}\,f, \mathrm{dmn}\,f)$ is a network. Throughout this Chapter we agree to let

$$S = \mathrm{dmn}\,f,$$
$$V = \sigma\,\mathrm{dmn}\,f = \sigma S.$$

Also, $\mathrm{rng}\,f$ will be significant in our work and it is convenient to let

$$K = \mathrm{rng}\,f.$$

We remind the reader that K is a set consisting of some, but not all ordered pairs (x, y) such that $\{x, y\}$ is an element of S. In fact, for each $\{x, y\} \in S$, there is precisely one of the ordered pairs (x, y) or (y, x) which is an element of K, and this unique ordered pair is equal to $f(\{x, y\})$. Thus, the incidence function f is univalent.

5.2 Chain Spaces

Consider $\mathscr{L}(K)$, the linear space of 0-chains of K. We shall also call $\mathscr{L}(K)$ the space of 1-chains of the set of branches S induced by the incidence function f.

If $u \in \mathscr{L}(K)$ and if $(x, y) \in K$, we shall always write

$$u(x, y)$$

instead of the notationally correct

$$u((x, y)).$$

We shall find it wortwhile to denote the zero element of $\mathscr{L}(K)$ by θ_1. Specifically, we let

$$\theta_1 = (K \times \{0\}),$$

and note that θ_1 is the function with domain K such that for each $(x, y) \in K$,

$$\theta_1(x, y) = 0 \in F.$$

Similarly, we shall denote the zero element of $\mathscr{L}(V)$ by θ_0. Thus, we let

$$\theta_0 = (V \times \{0\}),$$

and note that θ_0 is the function with domain V such that for each $x \in V$,

$$\theta_0(x) = 0 \in F.$$

5.3 The Boundary Operator

The boundary operator $\varDelta$ will be defined as a linear map of the space of 1-chains $\mathscr{L}(K)$ into the space of 0-chains $\mathscr{L}(V)$. Noting the Theorem of 4.6, it suffices to specify the element $\varDelta(\varPhi^K_{(x,y)})$ of $\mathscr{L}(K)$ for each element (x, y) of K, and we do this by specifying that $\varDelta(\varPhi^K_{(x,y)})$ should be equal to $\varPhi^V_y - \varPhi^V_x$ for each $(x, y) \in K$. Formally, we define,

$\varDelta$ is the unique linear map of $\mathscr{L}(K)$ into $\mathscr{L}(V)$ such that for each $(x, y) \in K$,
$$\varDelta(\varPhi^K_{(x,y)}) = \varPhi^V_y - \varPhi^V_x.$$

To justify the name "boundary operator" let $(x, y) \in K$ and consider the following.

Our convention is such that in a geometrical realization of the network (V, S) a directed branch from the vertex corresponding to x to the vertex corresponding to y corresponds to the ordered pair (x, y). The boundary of this branch consists of the vertices corresponding to x and y respectively, with an orientation assigned to these vertices consistent with the direction of the branch. Now, in $\mathscr{L}(K)$ the branch itself is identified with $\varPhi^K_{(x,y)}$, while in $\mathscr{L}(V)$ the boundary of the branch is identified with $\varPhi^V_y - \varPhi^V_x$, which is our definition of $\varDelta(\varPhi^K_{(x,y)})$.

5.4 Boundaries and Cycles

Let u be a linear map of A into B, and let θ be the zero element of B. Recall from elementary linear algebra that kernel u consists of all elements x of A such that $u(x)$ is equal to θ. Formally,

$$\text{kernel } u = \{x \mid x \in A \text{ and } u(x) = \theta\}.$$

From elementary linear algebra it is well known that kernel u is a subspace of A. It is equally well known that rng u is a subspace of B.

In the following Theorem we document these facts in the special case of our boundary operator Δ.

Theorem.

(1)   kernel Δ is a subspace of $\mathscr{L}(K)$.

(2)   rng Δ is a subspace of $\mathscr{L}(V)$.

The elements of kernel Δ will be called cycles. The elements of rng Δ will be called boundaries.

5.5 Summation Over Finite Sets

At this point we digress momentarily to discuss our notation involving summation over finite sets. We have made use of this notation previously, without comment, assuming that the reader could endow the notation with its proper interpretation. In this paragraph we make this notation more precise, in order to introduce some more elaborate assumptions involving this type of summation.

Let M be an additive abelian semi-group with additive identity element θ. Let u be any function such that rng $u \subset M$ and let A be a finite non-empty subset of dmn u. Previously, we have used the symbol

$$\sum_{x \in A} u(x)$$

with no further explanation. To define the above symbol precisely consider any z such that z is an ordering of A. Then, the above sum is defined to be equal to

$$\sum_{i=1}^{pA} u(z_i),$$

and its value is independent of the particular ordering z selected.

We now extend the above notation to cover the case when A is empty. In such a case the sum is defined to be the additive identity element of the semi-group. Thus, we define formally,

$$\sum_{x \in 0} u(x) = \theta.$$

We shall have occasion to consider products over finite sets and assume that the reader can extend the preceding summation notation to the proper notation involving products. We remind the reader that in such an extension the additive identity element must be replaced by the multiplicative identity element.

Next we need an assumption which will give us conditions sufficient to assert that two sums over different sets are equal. In particular, suppose that A and B are finite sets. Let s be a function with domain A and rng $s \subset M$ and let t be a function with domain B and rng $t \subset M$. We seek conditions sufficient to guarantee that

$$\sum_{x \in A} s(x) = \sum_{y \in B} t(y).$$

By using the preceding definition of our summation it can be shown that the above equation will be satisfied if we can establish the existence of a univalent function h with domain A and range B such that for each $x \in A$, $s(x) = t(h(x))$. We state this fact below as a formal Assumption.

Assumption. *Let M be an additive abelian semi-group with an identity element.*
Let s be a finite function such that rng $s \subset M$.
Let t be a finite function such that rng $t \subset M$.
Let h be a univalent function.
Suppose that dmn $h =$ dmn s *and* rng $h =$ dmn t *and* $s = t \circ h$.
Then,

$$\sum_{x \in \text{dmn } s} s(x) = \sum_{y \in \text{dmn } t} t(y).$$

5.6 The Coboundary Operator

The coboundary operator δ will be defined as a linear map of the space of 0-chains $\mathscr{L}(V)$ into the space of 1-chains $\mathscr{L}(K)$. From the Theorem of 4.6 it suffices to specify $\delta(\Phi_x^V)$ for each element x of V. To motivate our specification of $\delta(\Phi_x^V)$ for each element x of V we consider a geometrical realization of the network (V, S).

Let x be some fixed element of V and in the geometrical realization let x' be the vertex corresponding to x. Consider all directed branches in the geometrical realization which have x' as a vertex; some of these branches well be directed towards the vertex x' while the others will be directed away from the vertex x'.

Let A consist of all those elements y of V such that in the geometrical realization the branch corresponding to $\{x, y\}$ appears and is directed away from x'. Note that A consists of all elements y of V such that (x, y) is an element of K, and thus,

$$A = \operatorname{rng}\,(K|\{x\}).$$

Similarly, let B consist of all those elements y of V such that in the geometrical realization the branch corresponding to $\{x, y\}$ appears and is directed towards x'. Note that B consists of all elements y of V such that (y, x) is an element of K, and thus,

$$B = \operatorname{rng}\,((\operatorname{inv} K)|\{x\}).$$

Now, for each element y of A we identify the branch (x, y) of K with the element $\varPhi^K_{(x, y)}$ of $\mathscr{L}(K)$, and we form the sum

$$\sum_{y \in A} \varPhi^K_{(x, y)}.$$

Similarly, for each element y of B we identify the branch (y, x) of K with the element $\varPhi^K_{(y, x)}$ of $\mathscr{L}(K)$, and we form the sum

$$\sum_{y \in B} \varPhi^K_{(y, x)}.$$

For the coboundary $\delta(\varPhi^K_x)$ we shall specify the element

$$\sum_{y \in B} \varPhi^K_{(y, x)} - \sum_{y \in A} \varPhi^K_{(x, y)}$$

of $\mathscr{L}(K)$. Thus, the coboundary $\delta(\varPhi^K_x)$ selects, after appropriate identification, the sum of those branches directed towards x' diminished by the sum of those branches directed away from x'.

Formally, we define,

δ is the unique linear map of $\mathscr{L}(V)$ into $\mathscr{L}(K)$ such that for each $x \in V$,

$$\delta(\varPhi^K_x) = \sum_{y \,\in\, \operatorname{rng}\,((\operatorname{inv} K)|\{x\})} \varPhi^K_{(y, x)} - \sum_{y \,\in\, \operatorname{rng}\,(K|\{x\})} \varPhi^K_{(x, y)}.$$

5.7 Coboundaries and Cocycles

Analogous to 5.4 we document the facts that kernel δ is a subspace of $\mathscr{L}(V)$ and rng δ is a subspace of $\mathscr{L}(K)$.

Theorem.

(1)   kernel δ is a subspace of $\mathscr{L}(V)$.

(2)   rng δ is a subspace of $\mathscr{L}(K)$.

The elements of kernel δ will be called cocycles. The elements of rng δ will be called coboundaries.

5.8 Boundaries, Coboundaries, and Inner Products

Let $u \in \mathscr{L}(K)$ and let $s \in \mathscr{L}(V)$. In this paragraph as Theorem 2 we exhibit the important relationship

$$\langle \varDelta(u), s \rangle = \langle u, \delta(s) \rangle.$$

To establish this result we first consider in Theorem 1 the special case when u and s are elements of the respective canonical bases for $\mathscr{L}(K)$ and $\mathscr{L}(V)$.

Theorem 1. *Let $b \in V$. Let $(x, y) \in K$.*
Then,

$$\langle \varDelta(\varPhi^K_{(x,\,y)}), \varPhi^V_b \rangle = \langle \varPhi^K_{(x,\,y)}, \delta(\varPhi^V_b) \rangle.$$

Proof: By the definition of δ and properties of the inner product,

$$\langle \varPhi^K_{(x,\,y)}, \delta(\varPhi^V_b) \rangle = \langle \varPhi^K_{(x,\,y)}, \sum_{z\,\in\,\mathrm{rng}\,((\mathrm{inv}\,K)\,|\,\{b\})} \varPhi^K_{(z,\,b)} - \sum_{z\,\in\,\mathrm{rng}\,(K\,|\,\{b\})} \varPhi^K_{(b,\,z)} \rangle$$

$$= \langle \varPhi^K_{(x,\,y)}, \sum_{z\,\in\,\mathrm{rng}\,((\mathrm{inv}\,K)\,|\,\{b\})} \varPhi^K_{(z,\,b)} \rangle - \langle \varPhi^K_{(x,\,y)}, \sum_{z\,\in\,\mathrm{rng}\,(K\,|\,\{b\})} \varPhi^K_{(b,\,z)} \rangle$$

$$= \sum_{z\,\in\,\mathrm{rng}\,((\mathrm{inv}\,K\,|\,\{b\})} \langle \varPhi^K_{(x,\,y)}, \varPhi^K_{(z,\,b)} \rangle - \sum_{z\,\in\,\mathrm{rng}\,(K\,|\,\{b\})} \langle \varPhi^K_{(x,\,y)}, \varPhi^K_{(b,\,z)} \rangle.$$

Also, by the definition of $\varDelta$ and properties of the inner product,

$$\langle \varDelta(\varPhi^K_{(x,\,y)}), \varPhi^V_b \rangle = \langle \varPhi^V_y - \varPhi^V_x, \varPhi^V_b \rangle = \langle \varPhi^V_y, \varPhi^V_b \rangle - \langle \varPhi^V_x, \varPhi^V_b \rangle.$$

We consider three cases.

Case 1. $b \neq x$ and $b \neq y$.

Since $b \neq x$,

$$\langle \varPhi^K_{(x,\,y)}, \varPhi^K_{(b,\,z)} \rangle = 0$$

for each $z \in \mathrm{rng}\,(K\,|\,\{b\})$, while

$$\langle \varPhi^V_x, \varPhi^V_b \rangle = 0.$$

Also, since $b \neq y$,

$$\langle \varPhi^K_{(x,\,y)}, \varPhi^K_{(z,\,b)} \rangle = 0$$

for each $z \in \mathrm{rng}\,((\mathrm{inv}\,K)|\{b\})$, while

$$\langle \Phi_y^V, \Phi_b^V \rangle = 0.$$

Thus, in this case,

$$\langle \Delta(\Phi_{(x,\,y)}^K), \Phi_b^V \rangle = 0 = \langle \Phi_{(x,\,y)}^K, \delta(\Phi_b^V) \rangle.$$

Case 2. $b = x$.

In this case, since $(x, y) \in K$, $x \neq y$ and thus, $b \neq y$, implying that

$$\langle \Phi_{(x,\,y)}^K, \Phi_{(z,\,b)}^K \rangle = 0$$

for each $z \in \mathrm{rng}\,((\mathrm{inv}\,K)|\{b\})$, and

$$\langle \Phi_y^V, \Phi_b^V \rangle = 0.$$

On the other hand, since $b = x$, $(b, y) \in K$ and $y \in \mathrm{rng}\,(K|\{b\})$, implying that

$$- \sum_{z\, \in\, \mathrm{rng}\,(K|\{b\})} \langle \Phi_{(x,\,y)}^K, \Phi_{(b,\,z)}^K \rangle = - \langle \Phi_{(b,\,y)}^K, \Phi_{(b,\,y)}^K \rangle = -1.$$

Also, since $b = x$,

$$-\langle \Phi_x^V, \Phi_b^V \rangle = -1.$$

Thus, in this case,

$$\langle \Delta(\Phi_{(x,\,y)}^K), \Phi_b^V \rangle = -1 = \langle \Phi_{(x,\,y)}^K, \delta(\Phi_b^V) \rangle.$$

Case 3. $b = y$.

In this case, since $(x, y) \in K$, $x \neq y$ and thus, $b \neq x$, implying that

$$\langle \Phi_{(x,\,y)}^K, \Phi_{(b,\,z)}^K \rangle = 0$$

for each $z \in \mathrm{rng}\,(K|\{b\})$, and

$$\langle \Phi_x^V, \Phi_b^V \rangle = 0.$$

On the other hand, since $b = y$, $(x, b) \in K$ and $x \in \mathrm{rng}\,((\mathrm{inv}\,K)|\{b\})$, implying that

$$\sum_{z\, \in\, \mathrm{rng}\,((\mathrm{inv}\,K)|\{b\})} \langle \Phi_{(x,\,y)}^K, \Phi_{(z,\,b)}^K \rangle = \langle \Phi_{(x,\,b)}^K, \Phi_{(x,\,b)}^K \rangle = 1.$$

Also, since $b = y$,

$$\langle \Phi_y^V, \Phi_b^V \rangle = 1.$$

Thus, in this case,

$$\langle \Delta(\Phi_{(x,\,y)}^K), \Phi_b^V \rangle = 1 = \langle \Phi_{(x,\,y)}^K, \delta(\Phi_b^V) \rangle,$$

completing the proof.

Theorem 2. *Let $u \in \mathscr{L}(K)$. Let $s \in \mathscr{L}(V)$.*
Then,

$$\langle \Delta(u), s \rangle = \langle u, \delta(s) \rangle.$$

Proof: By the linearity of Δ, the linearity of δ, properties of the inner product, and Theorem 1,

$$\begin{aligned}
\langle \Delta(u), s \rangle &= \langle \Delta\big(\sum_{(x,y) \in K} u(x,y)\, \Phi^K_{(x,y)} \big), \sum_{b \in V} s(b)\, \Phi^V_b \rangle \\
&= \sum_{b \in V} s(b) \langle \sum_{(x,y) \in K} u(x,y)\, \Delta(\Phi^K_{(x,y)}), \Phi^V_b \rangle \\
&= \sum_{b \in V} \sum_{(x,y) \in K} u(x,y)\, s(b) \langle \Delta(\Phi^K_{(x,y)}), \Phi^V_b \rangle \\
&= \sum_{(x,y) \in K} \sum_{b \in V} u(x,y)\, s(b) \langle \Phi^K_{(x,y)}, \delta(\Phi^V_b) \rangle \\
&= \langle \sum_{(x,y) \in K} u(x,y)\, \Phi^K_{(x,y)}, \sum_{b \in V} s(b)\, \delta(\Phi^V_b) \rangle \\
&= \langle u, \delta\big(\sum_{b \in V} s(b)\, \Phi^V_b \big) \rangle \\
&= \langle u, \delta(s) \rangle.
\end{aligned}$$

5.9 Orthogonality of Cycles and Coboundaries

Observe that the space of cycles and the space of coboundaries are both subspaces of $\mathscr{L}(K)$. In the following Theorem 1 we show that the space of cycles is orthogonal to the space of coboundaries. By this we mean that the inner product of any cycle and any coboundary is equal to zero.

Theorem 1. *Let $u \in$ kernel Δ. Let $v \in$ rng δ.*
Then,
$$\langle u, v \rangle = 0.$$

Proof: Produce $s \in \mathscr{L}(V)$ such that $v = \delta(s)$. By Theorem 2 of 5.8,

$$\langle u, v \rangle = \langle u, \delta(s) \rangle = \langle \Delta(u), s \rangle.$$

But since $u \in$ kernel Δ, $\Delta(u) = \theta_0$.
Thus,

$$\langle u, v \rangle = \langle \Delta(u), s \rangle = \langle \theta_0, s \rangle = 0.$$

Since the set of cycles is a subspace of $\mathscr{L}(K)$, θ_1 is a cycle. Similarly, since the set of coboundaries is a subspace of $\mathscr{L}(K)$, θ_1 is a coboundary. Thus, θ_1 is both a cycle and a coboundary. In the following Corollary we point out that there are no other elements of $\mathscr{L}(K)$ wich are both cycles and coboundaries.

Corollary. kernel $\Delta \cap$ rng $\delta = \{\theta_1\}$.

Proof: Since trivially, $\{\theta_1\} \subset$ kernel $\Delta \cap$ rng δ, we shall show that kernel $\Delta \cap$ rng $\delta \subset \{\theta_1\}$.

Let $u \in$ kernel $\Delta \cap$ rng δ.
By Theorem 1, $\langle u, u \rangle = 0$.
Thus, $u = \theta_1$.

A partial converse to Theorem 1 is also valid. In particular, in the following Theorem 2 we show that any element of $\mathscr{L}(K)$ which is orthogonal to every coboundary must be a cycle.

Theorem 2. *Let $u \in \mathscr{L}(K)$. Suppose that*

$$v \in \text{rng } \delta \text{ implies } \langle u, v \rangle = 0.$$

Then,

$\quad u \in$ kernel Δ.

Proof: Observe that $\delta(\Delta(u)) \in$ rng δ.
Thus, applying the hypothesis to $\delta(\Delta(u))$,

$$\langle u, \delta(\Delta(u)) \rangle = 0.$$

But by Theorem 2 of 5.8,

$$0 = \langle u, \delta(\Delta(u)) \rangle = \langle \Delta(u), \Delta(u) \rangle.$$

Thus, $\Delta(u) = \theta_0$, implying that $u \in$ kernel Δ.

Later in this Chapter we shall establish another partial converse of Theorem 1 in which the roles played by coboundaries and cycles in Theorem 2 are interchanged.

5.10 Orthogonality of Boundaries and Cocycles

Observe that the space of boundaries and the space of cocycles are both subspaces of $\mathscr{L}(V)$. In the following Theorem 1 we show that the space of boundaries is orthogonal to the space of cocycles.

Theorem 1. *Let $s \in$ rng Δ. Let $t \in$ kernel δ.*
Then,

$\quad \langle s, t \rangle = 0$.

Proof: Produce $u \in \mathscr{L}(K)$ such that $s = \Delta(u)$. By Theorem 2 of 5.8,

$$\langle s, t \rangle = \langle \Delta(u), t \rangle = \langle u, \delta(t) \rangle.$$

But since $t \in$ kernel δ, $\delta(t) = \theta_1$.
Thus,

$$\langle s, t \rangle = \langle u, \delta(t) \rangle = \langle u, \theta_1 \rangle = 0.$$

Since the set of boundaries is a subspace of $\mathscr{L}(V)$, θ_0 is a boundary. Similarly, since the set of cocycles is a subspace of $\mathscr{L}(V)$, θ_0 is a cocycle. Thus, θ_0 is both a boundary and a cocycle. In the following Corollary we point out that there are no other elements of $\mathscr{L}(V)$ which are both boundaries and cocycles.

Corollary. kernel $\delta \cap$ rng $\varDelta = \{\theta_0\}$.

Proof: Since trivially, $\{\theta_0\} \subset$ kernel $\delta \cap$ rng $\varDelta$, we shall show that kernel $\delta \cap$ rng $\varDelta \subset \{\theta_0\}$.

Let $s \in$ kernel $\delta \cap$ rng $\varDelta$.
By Theorem 1, $\langle s, s \rangle = 0$.
Thus, $s = \theta_0$.

A partial converse to Theorem 1 is also valid. In particular, in the following Theorem 2 we show that any element of $\mathscr{L}(V)$ which is ortho-, gonal to every boundary must be a cocycle.

Theorem 2. *Let $s \in \mathscr{L}(V)$. Suppose that*

$$t \in \text{rng } \varDelta \text{ implies } \langle s, t \rangle = 0.$$

Then,

    $s \in$ kernel δ.

Proof: Observe that $\varDelta(\delta(s)) \in$ rng $\varDelta$. Applying the hypothesis to $\varDelta(\delta(s))$,

$$\langle s, \varDelta(\delta(s)) \rangle = \langle \varDelta(\delta(s)), s \rangle = 0.$$

But by Theorem 2 of 5.8,

$$0 = \langle \varDelta(\delta(s)), s \rangle = \langle \delta(s), \delta(s) \rangle.$$

Thus, $\delta(s) = \theta_1$, implying that $s \in$ kernel δ.

Later in this Chapter we shall establish another partial converse of Theorem 1 in which the roles played by boundaries and cocycles in Theorem 2 are interchanged.

5.11 Decomposition of $\mathscr{L}(K)$ into Cycles and Coboundaries

We know that the space of cycles is a subspace of $\mathscr{L}(K)$ and the space of coboundaries is a subspace of $\mathscr{L}(K)$. In the following Theorem we show that $\mathscr{L}(K)$ is actually decomposed into the direct sum of the space of cycles and the space of coboundaries.

Theorem. $\mathscr{L}(K) =$ kernel $\varDelta \oplus$ rng δ.

Proof: In view of the Theorem of 4.8 together with the Corollary of 5.9 it suffices to show that

$$\mathscr{L}(K) = \text{kernel } \varDelta + \text{rng } \delta.$$

Let $u \in \mathscr{L}(K)$.

We shall exhibit $v \in$ kernel Δ and $w \in$ rng δ such that $u = v + w$.

Now, if rng $\delta = \{\theta_1\}$, then Theorem 2 of 5.9 implies that $u \in$ kernel Δ, and thus, by letting $v = u$ and $w = \theta_1$, we have $u = v + w$, and we are through.

Hence, we may assume that rng $\delta \neq \{\theta_1\}$, or equivalently, dim (rng δ)>0.

By Theorem 4 of 4.5, produce B such that B is an orthonormal base of rng δ. Thus,

$$x \in B \text{ implies } \langle x, x \rangle = 1.$$
$$x \in B \text{ and } y \in B \text{ and } x \neq y$$
$$\text{implies}$$
$$\langle x, y \rangle = 0.$$

Now, let

$$w = \sum_{b \in B} \langle b, u \rangle \, b,$$

and observe that $w \in$ rng δ.

Let

$$v = u - w,$$

and observe that to complete the proof it suffices to show that $v \in$ kernel Δ. This is done in the remainder of the proof which is divided into three parts.

Part 1. $c \in B$ implies $\langle c, v \rangle = 0$.

Proof of Part 1: Let $c \in B$.

Then,

$$\begin{aligned}
\langle c, v \rangle = \langle c, u - w \rangle &= \langle c, u \rangle - \langle c, w \rangle \\
&= \langle c, u \rangle - \langle c, \sum_{b \in B} \langle b, u \rangle \, b \rangle \\
&= \langle c, u \rangle - \sum_{b \in B} \langle b, u \rangle \langle c, b \rangle \\
&= \langle c, u \rangle - \langle c, u \rangle \langle c, c \rangle \\
&= \langle c, u \rangle - \langle c, u \rangle = 0.
\end{aligned}$$

Part 2. $r \in$ rng δ implies $\langle v, r \rangle = 0$.

Proof of Part 2: Let $r \in$ rng δ.

By Theorem 5 of 4.5,

$$r = \sum_{c \in B} \langle c, r \rangle \, c.$$

By Part 1,

$$\langle v, r \rangle = \langle v, \sum_{c \in B} \langle c, r \rangle \, c \rangle = \sum_{c \in B} \langle c, r \rangle \langle v, c \rangle = 0.$$

Part 3. $v \in$ kernel Δ.

Proof of Part 3: Immediate, by Part 2 and Theorem 2 of 5.9.

As a Corollary to the preceding Theorem we are now able to exhibit the second partial converse of Theorem 1 of 5.9 as advertised in the discussion of 5.9.

Corollary. *Let* $u \in \mathscr{L}(K)$. *Suppose that*

$$v \in \text{kernel } \Delta \ \textit{implies} \ \langle u, v \rangle = 0.$$

Then,

$u \in \text{rng } \delta.$

Proof: By the preceding Theorem, produce $v \in \text{kernel } \Delta$ and $w \in \text{rng } \delta$ such that $u = v + w$.
Then, by the hypothesis,

$$0 = \langle u, v \rangle = \langle v + w, v \rangle = \langle v, v \rangle + \langle w, v \rangle.$$

But by Theorem 1 of 5.9, $\langle w, v \rangle = 0$.
Thus, $0 = \langle v, v \rangle$, and $v = \theta_1$, implying that

$$u = \theta_1 + w = w \in \text{rng } \delta.$$

5.12 Decomposition of $\mathscr{L}(V)$ into Boundaries and Cocycles

We know that the space of boundaries is a subspace of $\mathscr{L}(V)$ and the space of cocycles is a subspace of $\mathscr{L}(V)$. In the following Theorem we show that $\mathscr{L}(V)$ is actually decomposed into the direct sum of the space of boundaries and the space of cocycles.

Theorem. $\mathscr{L}(V) = \text{kernel } \delta \oplus \text{rng } \Delta$.

Proof: In view of the Theorem of 4.8 together with the Corollary of 5.10 it suffices to show that

$$\mathscr{L}(V) = \text{kernel } \delta + \text{rng } \Delta.$$

Let $s \in \mathscr{L}(V)$.
We shall exhibit $t \in \text{kernel } \delta$ and $m \in \text{rng } \Delta$ such that $s = t + m$.

Now, if $\text{rng } \Delta = \{\theta_0\}$, then Theorem 2 of 5.10 implies that $s \in \text{kernel } \delta$, and thus, by letting $t = s$ and $m = \theta_0$, we have $s = t + m$, and we are through.
Hence, we may assume that $\text{rng } \Delta \neq \{\theta_0\}$, or equivalently, $\dim (\text{rng } \Delta) > 0$.

By Theorem 4 of 4.5, produce B such that B is an orthonormal base of $\text{rng } \Delta$. Thus,

$$x \in B \text{ implies } \langle x, x \rangle = 1.$$
$$x \in B \text{ and } y \in B \text{ and } x \neq y$$
$$\text{implies}$$
$$\langle x, y \rangle = 0.$$

7*

Now, let

$$m = \sum_{b \in B} \langle b, s \rangle \, b,$$

and observe that $m \in \mathrm{rng}\ \varDelta$.

Let

$$t = s - m,$$

and observe that to complete the proof it suffices to show that $t \in \mathrm{kernel}\ \delta$. This is done in the remainder of the proof which is divided into three parts.

Part 1. $c \in B$ implies $\langle c, t \rangle = 0$.

Proof of Part 1: Let $c \in B$. Then,

$$\begin{aligned}
\langle c, t \rangle = \langle c, s - m \rangle &= \langle c, s \rangle - \langle c, m \rangle \\
&= \langle c, s \rangle - \langle c, \sum_{b \in B} \langle b, s \rangle \, b \rangle \\
&= \langle c, s \rangle - \sum_{b \in B} \langle b, s \rangle \langle c, b \rangle \\
&= \langle c, s \rangle - \langle c, s \rangle \langle c, c \rangle \\
&= \langle c, s \rangle - \langle c, s \rangle = 0.
\end{aligned}$$

Part 2. $r \in \mathrm{rng}\ \varDelta$ implies $\langle t, r \rangle = 0$.

Proof of Part 2: Let $r \in \mathrm{rng}\ \varDelta$.

By Theorem 5 of 4.5,

$$r = \sum_{c \in B} \langle c, r \rangle \, c.$$

By Part 1,

$$\langle t, r \rangle = \langle t, \sum_{c \in B} \langle c, r \rangle \, c \rangle = \sum_{c \in B} \langle c, r \rangle \langle t, c \rangle = 0.$$

Part 3. $t \in \mathrm{kernel}\ \delta$.

Proof of Part 3: Immediate, by Part 2 and Theorem 2 of 5.10.

The reader should note the duality between the proof of the preceding Theorem and the proof of the Theorem of 5.11. We have deliberately included all of the tedious details of the preceding proof in order to display this duality completely.

As a Corollary to the preceding Theorem we are now able to exhibit the second partial converse of Theorem 1 of 5.10 as advertised in the discussion of 5.10. To exhibit the duality with the Corollary of 5.11 we again display all of the tedious details.

Corollary. *Let $s \in \mathscr{L}(V)$. Suppose that*

$$t \in \mathrm{kernel}\ \delta \ \ implies \ \ \langle s, t \rangle = 0.$$

Then,

$s \in \operatorname{rng} \Delta$.

Proof: By the preceding Theorem, produce $t \in \operatorname{kernel} \delta$ and $m \in \operatorname{rng} \Delta$ such that $s = t + m$.

Then, by the hypothesis,

$$0 = \langle s, t \rangle = \langle t + m, t \rangle = \langle t, t \rangle + \langle m, t \rangle.$$

But by Theorem 1 of 5.10, $\langle m, t \rangle = 0$.

Thus, $0 = \langle t, t \rangle$, and $t = \theta_0$, implying that

$$s = \theta_0 + m = m \in \operatorname{rng} \Delta.$$

5.13 Isomorphism of Coboundaries and Boundaries

The space of coboundaries is isomorphic to the space of boundaries. By this, we mean that there exists a linear map of the space of coboundaries into the space of boundaries which is univalent and whose range is equal to the space of boundaries. In the following Theorem we establish the isomorphism of the space of coboundaries and the space of boundaries, and show that the restriction of the boundary operator to the space of coboundaries is a linear map meeting the above specifications.

Theorem. *rng δ is isomorphic to rng Δ.*

Proof: Let $g = \Delta | \operatorname{rng} \delta$.

The linearity of Δ guarantees that g is a linear map of $\operatorname{rng} \delta$ into $\operatorname{rng} \Delta$. Thus, to complete the proof it suffices to show that $\operatorname{rng} g = \operatorname{rng} \Delta$ and g is univalent. This is done in the remainder of the proof which is divided into two parts.

Part 1. $\operatorname{rng} g = \operatorname{rng} \Delta$.

Proof of Part 1: It suffices to show that $\operatorname{rng} \Delta \subset \operatorname{rng} g$.

Let $x \in \operatorname{rng} \Delta$.

Produce $u \in \mathscr{L}(K)$ such that $x = \Delta(u)$.

By the Theorem of 5.11, produce $v \in \operatorname{kernel} \Delta$ and $w \in \operatorname{rng} \delta$ such that $u = v + w$.

Then,

$$x = \Delta(u) = \Delta(v + w) = \Delta(v) + \Delta(w) = \theta_0 + \Delta(w) = \Delta(w).$$

But since $w \in \operatorname{rng} \delta$,

$$x \in \operatorname{rng} (\Delta | \operatorname{rng} \delta) = \operatorname{rng} g,$$

completing the proof of Part 1.

Part 2. g is univalent.

Proof of Part 2: Let $u \in \mathrm{dmn}\ g$ and let $v \in \mathrm{dmn}\ g$ such that $g(u) = g(v)$.

Since $g = \Delta|\mathrm{rng}\ \delta$, we have

$$u \in \mathrm{rng}\ \delta,$$
$$v \in \mathrm{rng}\ \delta,$$
$$\Delta(u) = \Delta(v).$$

Since $\mathrm{rng}\ \delta$ is a subspace of $\mathscr{L}(K)$, $u - v \in \mathrm{rng}\ \delta$.

Also, since $\Delta(u) = \Delta(v)$,

$$\theta_0 = \Delta(u) - \Delta(v) = \Delta(u - v),$$

implying that $u - v \in \mathrm{kernel}\ \Delta$.

Thus, by the Corollary of 5.9,

$$u - v \in \mathrm{kernel}\ \Delta \cap \mathrm{rng}\ \delta = \{\theta_1\},$$

implying that $u - v = \theta_1$, and finally, $u = v$, completing the proof.

Form elementary linear algebra we know that two finite dimensional linear spaces which are isomorphic have the same dimension. Thus, the dimension of the space of coboundaries is equal to the dimension of the space of boundaries. This fact is stated in the following Corollary 1, but we omit the proof.

Corollary 1. $\dim(\mathrm{rng}\ \Delta) = \dim(\mathrm{rng}\ \delta)$.

We can combine our previous results to establish a fundamental result involving the dimensions of some of the spaces under discussion. In particular, in the following Corollary 2 we show that the dimension of $\mathscr{L}(K)$ plus the dimension of the space of cocycles is equal to the dimension of $\mathscr{L}(V)$ plus the dimension of the space of cycles.

Corollary 2.
$$\dim(\mathscr{L}(K)) + \dim(\mathrm{kernel}\ \delta) = \dim(\mathscr{L}(V)) + \dim(\mathrm{kernel}\ \Delta).$$

Proof: By the Theorem of 4.9 and the Theorem of 5.11,

$$\dim(\mathscr{L}(K)) = \dim(\mathrm{kernel}\ \Delta) + \dim(\mathrm{rng}\ \delta).$$

By the Theorem of 4.9 and the Theorem of 5.12,

$$\dim(\mathscr{L}(V)) = \dim(\mathrm{kernel}\ \delta) + \dim(\mathrm{rng}\ \Delta).$$

Thus, using Corollary 1 of this paragraph,

$$\dim(\mathscr{L}(K)) + \dim(\text{kernel } \delta)$$
$$= \dim(\text{kernel } \Delta) + \dim(\text{rng } \delta) + \dim(\text{kernel } \delta)$$
$$= \dim(\text{kernel } \Delta) + \dim(\text{kernel } \delta) + \dim(\text{rng } \Delta)$$
$$= \dim(\text{kernel } \Delta) + \dim(\mathscr{L}(V)).$$

5.14 Dimension of the Space of Cocycles

In this paragraph we investigate the dimension of the space of cocycles. Under the additional assumption that the network (V, S) is connected we show that the dimension of the space of cocycles is equal to 1. This is accomplished by proving that

$$\{\sum_{b \in V} \Phi_b^V\}$$

is a base of the space of cocycles.

This program requires some work. In Theorem 1 we show that $\sum_{b \in V} \Phi_b^V \neq \theta_0$ while in Theorem 2 we show that $\sum_{b \in V} \Phi_b^V$ is indeed a cocycle. We do not need the connectedness of (V, S) to establish these first two theorems. Finally, in Theorem 3 we show that under the assumption of the connectedness of (V, S), $\{\sum_{b \in V} \Phi_b^V\}$ is a base of the space of cocycles.

Theorem 1.

$$\sum_{b \in V} \Phi_b^V \neq \theta_0.$$

Proof: Since $\sum_{b \in V} \Phi_b^V$ is a function with domain V, we let $c \in V$ and show that $\left(\sum_{b \in V} \Phi_b^V\right)(c) \neq 0$.

$$\left(\sum_{b \in V} \Phi_b^V\right)(c) = \sum_{b \in V} \Phi_b^V(c) = \Phi_c^V(c) = 1 \neq 0.$$

Theorem 2.

$$\sum_{b \in V} \Phi_b^V \in \text{kernel } \delta.$$

Proof: We show that $\delta\left(\sum_{b \in V} \Phi_b^V\right) = \theta_1$.

Since $\delta\left(\sum_{b \in V} \Phi_b^V\right)$ is a function with domain K, we let $(s, t) \in K$. It suffices to show that $\left(\delta\left(\sum_{b \in V} \Phi_b^V\right)\right)(s, t) = 0$.

$$\left(\delta\left(\sum_{b \in V}\Phi_b^V\right)\right)(s,t) = \left(\sum_{b \in V}\delta(\Phi_b^V)\right)(s,t)$$

$$= \left(\sum_{b \in V}\left(\sum_{y \in \mathrm{rng}\,((\mathrm{inv}\,K)\,|\,\{b\})}\Phi_{(y,\,b)}^K - \sum_{y \in \mathrm{rng}\,(K\,|\,\{b\})}\Phi_{(b,\,y)}^K\right)\right)(s,t)$$

$$= \left(\sum_{b \in V}\sum_{y \in \mathrm{rng}\,((\mathrm{inv}\,K)\,|\,\{b\})}\Phi_{(y,\,b)}^K - \sum_{b \in V}\sum_{y \in \mathrm{rng}\,(K\,|\,\{b\})}\Phi_{(b,\,y)}^K\right)(s,t)$$

$$= \sum_{b \in V}\sum_{y \in \mathrm{rng}\,((\mathrm{inv}\,K)\,|\,\{b\})}\Phi_{(b,\,y)}^K(s,t) - \sum_{b \in V}\sum_{y \in \mathrm{rng}\,(K\,|\,\{b\})}\Phi_{(b,\,y)}^K(s,t)$$

$$= \sum_{y \in \mathrm{rng}\,((\mathrm{inv}\,K)\,|\,\{t\})}\Phi_{(y,\,t)}^K(s,t) - \sum_{y \in \mathrm{rng}\,(K\,|\,\{s\})}\Phi_{(s,\,y)}^K(s,t)$$

$$= \Phi_{(s,\,t)}^K(s,t) - \Phi_{(s,\,t)}^K(s,t) = 0.$$

Theorem 3. *Suppose that (V,S) is a connected network. Then,*

dim (kernel δ) $= 1$.

Proof: By Theorem 2 of this paragraph,

$$\left\{\sum_{b \in V}\Phi_b^V\right\} \subset \text{kernel } \delta.$$

Thus, it suffices to show that $\left\{\sum_{b \in V}\Phi_b^V\right\}$ is a base of kernel δ.

By Theorem 1 of this paragraph $\left\{\sum_{b \in V}\Phi_b^V\right\}$ is linearly independent. Thus, it suffices to show that $\left\{\sum_{b \in V}\Phi_b^V\right\}$ generates kernel δ.

Let $s \in$ kernel δ. It suffices to exhibit $x \in F$ such that $s = x\sum_{b \in V}\Phi_b^V$.

Now, $s \in \mathscr{L}(V)$ and $\{\Phi_b^V\,|\,b \in V\}$ is a base of $\mathscr{L}(V)$. Thus, produce a function g with dmn $g = V$ and rng $g \subset F$ such that

$$s = \sum_{b \in V}g(b)\,\Phi_b^V.$$

To complete the proof it suffices to show that g is a constant function. More precisely, we shall show that $g(a) = g(c)$ whenever $a \in V$ and $c \in V$ and $a \neq c$.

Let $a \in V$ and let $c \in V$ with $a \neq c$. In the remainder of the proof which is divided into two parts we show that $g(a) = g(c)$.

Part 1. $(v,t) \in K$ implies $g(v) = g(t)$.

Proof of Part 1: Let $(v,t) \in K$.

Since $s \in$ kernel δ, $\delta(s) = \theta_1$, and $(\delta(s))(v,t) = 0$.

Thus,

$$
\begin{aligned}
0 &= (\delta(s))\,(v,t) \\
&= \left(\delta\left(\sum_{b\in V} g(b)\,\Phi_b^V\right)\right)(v,t) \\
&= \left(\sum_{b\in V} g(b)\,\delta(\Phi_b^V)\right)(v,t) \\
&= \left(\sum_{b\in V} g(b)\left(\sum_{y\in \operatorname{rng}((\operatorname{inv}K)\,|\,\{b\})} \Phi_{(y,b)}^K \right.\right. \\
&\qquad\quad \left.\left. -\sum_{y\in \operatorname{rng}(K\,|\,\{b\})} \Phi_{(b,y)}^K\right)\right)(v,t) \\
&= \left(\sum_{b\in V}\ \sum_{y\in \operatorname{rng}((\operatorname{inv}K)\,|\,\{b\})} g(b)\,\Phi_{(y,b)}^K \right. \\
&\qquad\quad \left. -\sum_{b\in V}\ \sum_{y\in \operatorname{rng}(K\,|\,\{b\})} g(b)\,\Phi_{(b,y)}^K\right)(v,t) \\
&= \sum_{b\in V}\ \sum_{y\in \operatorname{rng}((\operatorname{inv}K)\,|\,\{b\})} g(b)\,\Phi_{(y,b)}^K\,(v,t) \\
&\qquad -\sum_{b\in V}\ \sum_{y\in \operatorname{rng}(K\,|\,\{b\})} g(b)\,\Phi_{(b,y)}^K\,(v,t) \\
&= \sum_{y\in \operatorname{rng}((\operatorname{inv}K)\,|\,\{t\})} g(t)\,\Phi_{(y,t)}^K\,(v,t) \\
&\qquad -\sum_{y\in \operatorname{rng}(K\,|\,\{v\})} g(v)\,\Phi_{(v,y)}^K\,(v,t) \\
&= g(t)\,\Phi_{(v,t)}^K\,(v,t) - g(v)\,\Phi_{(v,t)}^K\,(v,t) \\
&= g(t) - g(v).
\end{aligned}
$$

Thus, $g(t) = g(v)$, completing the proof of Part 1.

Part 2. $g(a) = g(c)$.

Proof of Part 2:

Since (V,S) is a connected network, S is path connected by the Theorem of 1.20. Thus, produce z such that z is a path from a to c in S.

Let $m \in \omega$ such that $\operatorname{dmn} z = [m]$.

Observe that $a \in z_1$ and $c \in z_m$.

For each $k \in [m]$, let $P(k)$ be the proposition

$$r \in z_k \text{ implies } g(a) = g(r).$$

To complete the proof we shall prove by mathematical induction that $P(k)$ is true for each $k \in [m]$.

To prove that $P(1)$ is true let $r \in z_1$. If $r = a$, then $g(a) = g(r)$, so we may assume that $r \neq a$, implying that $z_1 = \{a, r\} \in S$. Thus, $(a, r) \in K$ or $(r, a) \in K$, and in either case, by Part 1, $g(a) = g(r)$.

Now, assume that $k \in [m]$ and $k + 1 \in [m]$ such that $P(k)$ is true. We must show that $P(k+1)$ is true.

Thus, let $y \in z_{k+1}$. We must show that $g(a) = g(y)$.
But since z is a path from a to c in S, $z_k \cap z_{k+1} \neq 0$.
Produce $n \in z_k \cap z_{k+1}$.
Since $n \in z_k$, the inductive hypothesis guarantees that $g(a) = g(n)$.

But $n \in z_{k+1}$.
If $n = y$, then $g(a) = g(n) = g(y)$, so we may assume that $n \neq y$,
implying that $z_{k+1} = \{y, n\} \in S$.
Thus, $(y, n) \in K$ or $(n, y) \in K$, and in either case, by Part 1,
$g(a) = g(n) = g(y)$, completing the proof.

As a Corollary we specialize the equation of Corollary 2 of 5.13 to
the case when (V, S) is a connected network.

Corollary. *Suppose that (V, S) is a connected network.*
Then,

$$\dim(\mathscr{L}(K)) + 1 = \dim(\mathscr{L}(V)) + \dim(\text{kernel } \Delta).$$

Proof: Immediate, by Corollary 2 of 5.13 and Theorem 3 of this
paragraph.

CHAPTER SIX

Axioms of Network Analysis

6.0 Introduction

In this Chapter we relate the mathematical abstractions of the preceding pages to the realities exemplified by an actual resistive network. We manufacture such a resistive network by assigning a positive real number to each branch of an abstract network. The resistive network, however, is not an isolated object; it participates in a more complicated circuit, and its behavior is described by its branch currents and branch voltages.

The branch currents and branch voltages must be related in such a way that three fundamental laws are satisfied. The first law is Ohm's Law, while the other two are Kirchhoff's Current Law and Kirchhoff's Voltage Law. In this Chapter we formulate these three laws precisely in terms of the mathematical abstractions previously introduced. Such a formulation provides a firm foundation upon which electrical circuit theory can be developed.

6.1 Assumptions of This Chapter

At this point in our development we can no longer work with an arbitrary field of scalars in our linear algebra. Hence, we agree that the field of scalars will be R, the field of real numbers.

As in Chapter 5 we fix an incidence function f, and agree to let

$$S = \operatorname{dmn} f,$$
$$V = \sigma \operatorname{dmn} f = \sigma S,$$
$$K = \operatorname{rng} f.$$

6.2 Resistive Networks

To transform a geometrical realization of our abstract network (V, S) into an actual resistive network we must assign a positive number to each branch of S. Such an assignment specifies a function, and in this

Chapter we shall consider one such function to be denoted by r. Thus, we agree to fix r such that

$$r \text{ is a function,}$$
$$\operatorname{dmn} r = S,$$
$$\operatorname{rng} r \subset \{x \mid x \in R \text{ and } x > 0\}.$$

For each branch $\{x, y\}$ of S we call $r(\{x, y\})$ the resistance of the branch $\{x, y\}$. Observe that for each branch $\{x, y\}$ of S, the resistance $r(\{x, y\})$ is independent of the direction assigned to the branch $\{x, y\}$ by the incidence function f.

Consider any geometrical realization of the network (V, S). By drawing the customary wiggly line in place of each branch of the geometrical realization, and then assigning to each wiggly line the value of the resistance of the branch, we obtain a circuit diagram of the resistive network. In fact, with an adequate supply of resistors we can manufacture an actual resistive network corresponding to our abstract network (V, S) and the resistance function r.

We point out that in this book we are only considering the circuit theory of resistive networks. To extend our theory to passive networks with capacitive and inductive elements we would assign a complex valued rational function with real coefficients to each branch of the network, and change the field of scalars to the field of complex numbers. A further extension of the theory to active networks would require more elaborate functions assigned to each branch of the network. Such extensions, of course, complicate the entire theory, and lead to a presentation much more involved than the theory of resistive networks which follows. Furthermore, the author feels that the topological properties of the networks can most clearly be exhibited when the theory is restricted to resistive networks. Introduction of more elaborate network elements would only abscure the basic topological principles which we are trying to exhibit in this book.

6.3 Currents and Voltages

Consider an actual resistive network manufactured from (V, S) and r. In real life this resistive network is not used alone but appears as a constituent part in a larger, more complicated circuit. From its placement in the larger, more complicated circuit our resistive network is endowed with electrical energy, and the circuit designer wants a means of describing the way in which the resistive network will participate in the larger, more complicated circuit. A description of the participation, or performance, or the resistive network is obtained by means of currents and voltages.

By a current in a directed branch of K we mean a real number assigned to this directed branch of K. Such a real number is usually called the current flow in this directed branch of K. By assigning a current which is a real number to each directed branch of K we manufacture a chain of $\mathscr{L}(K)$. Thus, we shall say that a current i is actually an element of $\mathscr{L}(K)$, and for any directed branch $(x, y) \in K$, $i(x, y)$ is the current flow in the branch (x, y).

By a voltage in a directed branch of K we mean a real number assigned to this directed branch of K. Such a real number is usually called the voltage drop across this directed branch of K. By assigning a voltage which is a real number to each branch of K we manufacture a chain of $\mathscr{L}(K)$. Thus, we shall say that a voltage v is actually an element of $\mathscr{L}(K)$, and for any directed branch $(x, y) \in K$, $v(x, y)$ is the voltage drop across the branch (x, y).

We see that when the resistive network is inserted into a larger, more complicated circuit, and endowed with electrical energy, a complete description of the performance of the network is obtained by knowledge of the resulting current chain and voltage chain. In the remainder of this book we shall be concerned with the determination of the current chain and voltage chain after the initial endowment of energy to the network is specified.

However, the current chain and voltage chain are not arbitrary elements of $\mathscr{L}(K)$. Physical considerations yield certain laws which the current chain and voltage chain must satisfy. In this Chapter we shall quote these laws in terms of the abstract setting which we have developed.

In the preceding paragraphs we have spoken rather loosely about electrical energy which is supplied to our resistive network as a result of its placement in the larger, more complicated circuit. This concept, involving the means by which electrical energy is supplied to our resistive network, must also be made more precise. We do this later in this Chapter with the introduction of objects called sources.

6.4 Ohm's Law

Consider an actual resistive network manufactured from the network (V, S) and the resistance function r, and insert this resistive network into a larger, more complicated circuit in such a way that the resistive network is endowed with electrical energy. From elementary electrical engineering principles we know that a current chain i of $\mathscr{L}(K)$ is developed and a voltage chain v of $\mathscr{L}(K)$ is developed, and i and v jointly describe the performance of our resistive network. Physical considerations guarantee the existence and uniqueness of the chains i and v, but we shall return to the problems of existence and uniqueness in the next Chapter. Here, we temporarily make the assumption of the existence and uniqueness of i and v.

The first relation which i and v must satisfy is Ohm's Law. Ohm's Law states that for each branch (x, y) of K,

$$v(x, y) = r(\{x, y\})\, i(x, y).$$

For our later work it is convenient to state Ohm's Law in a less transparent, but more elegant way. For this purpose we agree, for the remainder of this Chapter to fix a function Z such that

Z is the unique linear map of $\mathscr{L}(K)$ into $\mathscr{L}(K)$
such that for each $(x, y) \in K$,
$$Z(\Phi^K_{(x,y)}) = r(\{x, y\})\, \Phi^K_{(x,y)}.$$

The function Z is called the impedance function of the network. Note that the incidence function f and the resistance function r are involved in the definition of Z. In particular, the incidence function f determines the domain of Z, and for each $(x, y) \in K, Z(\Phi^K_{(x,y)})$ depends upon the resistance of the branch $\{x, y\}$.

By means of the impedance function Z, Ohm's Law takes the simple formulation
$$v = Z(i).$$

To see that the above formulation of Ohm's Law is indeed correct, consider any $(x, y) \in K$.

Then, since $i = \sum_{(s,t)\in K} i(s, t)\, \Phi^K_{(s,t)}$,

$$(Z(i))\,(x, y) = \Big(Z\big(\sum_{(s,t)\in K} i(s, t)\, \Phi^K_{(s,t)}\big)\Big)\,(x, y)$$

$$= \Big(\sum_{(s,t)\in K} i(s, t)\, Z(\Phi^K_{s,t})\Big)\,(x, y)$$

$$= \Big(\sum_{(s,t)\in K} i(s, t)\, r(\{s, t\})\, \Phi^K_{(s,t)}\Big)\,(x, y)$$

$$= \sum_{(s,t)\in K} i(s, t)\, r(\{s, t\})\, \Phi^K_{(s,t)}\,(x, y)$$

$$= i(x, y)\, r(\{x, y\})\, \Phi^K_{(x,y)}\,(x, y)$$

$$= i(x, y)\, r(\{x, y\}).$$

6.5 Sources

Consider an actual resistive network manufactured from the network (V, S) and the resistance function r, and insert this resistive network into a larger, more complicated circuit in such a way that the resistive network is endowed with electrical energy. In this paragraph we define objects called sources in such a way that the performance of the isolated resistive network plus the sources will be equivalent to the performance of the resistive network when it is a constituent part of the larger, more complicated circuit. More precisely, these sources will induce the same current chain and the same voltage chain describing the performance of the resistive network, which would result from the insertion of the resistive network into the larger, more complicated circuit.

We assume that the resistive network, when inserted into the larger, more complicated circuit, receives its electrical energy in such a way that the energy at each point of introduction is introduced into a branch of the resistive network. In engineering practice, any insertion of the resistive network into a larger, more complicated circuit can always be made equivalent to a similar insertion in which the electrical energy is introduced into the branches of the resistive network. Thus, we lose no generality by this assumption.

The energy introduced into the branches of the resistive network when the resistive network is inserted into the larger, more complicated circuit may be in the form of a current or a voltage. Thus, we shall consider current sources and voltage sources separately. There are advanced theorems in circuit theory pointing out that the effect of one type of source upon the resistive network is equivalent to an appropriately selected source of the other type, but we shall not concern ourselves with such advanced theorems in this book. However, because of these advanced theorems stipulating the equivalence of the two types of sources, we may assume that either all of the sources are current sources or all of the sources are voltage sources.

We consider first the case of a network energized with a collection of current sources. Each such current source introduces a current into the network by selecting one branch of the network and injecting current into one vertex of this branch and removing the same current from the other vertex of this branch. Thus, each such current source assigns a number to this selected branch equal to the value of the injected external current flow into one vertex of this branch and out of the other vertex of this branch. Note, however, that this injected external current flow into the network has a direction as well as a magnitude. The direction of the injected external current flow is determined in the following way.

Suppose that the selected branch is $\{x, y\}$ and $f(\{x, y\}) = (x, y)$. If the value of the current source is positive, then the injected external current is introduced into the vertex x and removed at the vertex y. If the value of the current source is negative, then the injected external current is introduced into the vertex y and removed at the vertex x.

In this way, the collection of current sources specifies a function whose domain is equal to K, the set of directed branches, and whose range consists of real numbers. Such a function is merely a chain of $\mathscr{L}(K)$, and thus, we conclude that a collection of current sources is an element of $\mathscr{L}(K)$. We call such a collection of current sources a current source chain. Formally,

A current source chain is an element of $\mathscr{L}(K)$.

The situation involving a collection of voltage sources is similar. Each such voltage source injects a voltage into a branch of the network, and the injected voltage is usually referred to as an injected voltage drop in the branch. Thus, we can say that each voltage source assigns to the branch into which it is injected a number equal to the value of the source. Note, however, that the injected voltage drop has a direction as well as a magnitude. Because of the usual direction of a current flow produced by a voltage drop, we assign a direction to the injected voltage drop in the following way.

If the value of the voltage source is negative, then the direction of the injected voltage drop coincides with the direction assigned to the branch by the incidence function f. If the value of the voltage source is positive, then the injected voltage drop is directed opposite to the direction assigned to the branch by the incidence function f.

In this way, the collection of voltage sources specifies a function whose domain is equal to K, the set of directed branches, and whose range consists of real numbers. Such a function is merely a chain of $\mathscr{L}(K)$, and thus, we conclude that a collection of voltage sources is an element of $\mathscr{L}(K)$. We call such a collection of voltage sources a voltage source chain. Formally,

A voltage source chain is an element of $\mathscr{L}(K)$.

6.6 Kirchhoff's Laws for Voltage Sources

Manufacture an actual resistive network from the network (V, S) and the resistance function r, and suppose that the resistive network is inserted into a larger, more complicated circuit in such a way that all energy introduced into the branches of the resistive network is in the form of voltages. Physical considerations assure us that these external voltages give rise to branch currents and branch voltages in the branches of the resistive network, describing the performance of the resistive network. We have pointed out that these branch currents and branch voltages must satisfy Ohm's Law. From elementary electrical engineering principles we also know that these branch currents and branch voltages must also satisfy two other laws due to Kirchhoff, Kirchhoff's Voltage Law and Kirchhoff's Current Law.

As stated in elementary electrical engineering, Kirchhoff's Voltage Law, in the presence of external voltages only, demands that the sum of the voltage drops along any closed path must be equal to the sum of the external voltages introduced into those branches comprising the closed path. As stated in elementary electrical engineering, Kirchhoff's Current Law, in the presence of external voltages only, demands that the total current flow into any vertex of the resistive network must be equal to zero, since no external currents are introduced into any branch which is incident to that vertex. In this Paragraph we translate these two laws of Kirchhoff into a precise formulation consistent with our abstract setting.

To make matters precise, suppose that the voltage source chain is the element e of $\mathscr{L}(K)$. Let $i \in \mathscr{L}(K)$ be the current chain induced by e and let $v \in \mathscr{L}(K)$ be the voltage chain induced by e as described in 6.3. We remind the reader that we are assuming the existence of such an $i \in \mathscr{L}(K)$ and $v \in \mathscr{L}(K)$ satisfying Ohm's Law. For each directed branch (x, y) of K, $i(x, y)$ is the value of the branch current and $v(x, y)$ is the value of the branch voltage.

In our formulation, the precise analogue of a closed path is a cycle. Thus, consider any cycle $u \in$ kernel $\varDelta$. We must translate the phrase "the sum of the voltage drops along any closed path" to a precise statement involving the branch voltage chain v and the particular cycle u.

At each directed branch (x, y) of K, $v(x, y)$ is the value of the branch voltage, but we cannot neglect the number $u(x, y)$ which the cycle u assigns to the branch (x, y). Loosely speaking, the number $u(x, y)$ measures the number of times this branch (x, y) is traversed and the direction of this travel during the cycle. The total contribution of the directed branch (x, y) to the precise analogue of the sum of the voltage drops along the closed path corresponding to the cycle u is the product,

$$v(x, y)\, u(x, y).$$

Thus, the precise analogue of the sum of the voltage drops along the entire closed path corresponding to the cycle u is the sum,

$$\sum_{(x,y)\,\in\,K} v(x, y)\, u(x, y),$$

which is equal to the inner product,

$$\langle v, u \rangle.$$

Similarly, the precise formulation of the total external voltage introduced into the branches of the closed path corresponding to u is the inner product,

$$\langle e, u \rangle.$$

Thus, Kirchhoff's Voltage Law in the presence of voltage sources requires that for each cycle $u \in$ kernel $\varDelta$,

$$\langle v, u \rangle = \langle e, u \rangle,$$

which is equivalent to the statement that

$$\langle v - e, u \rangle = 0,$$

for each cycle $u \in$ kernel $\varDelta$.

But by the Corollary of 5.11 the preceding statement implies that $v - e$ is a coboundary. By Theorem 1 of 5.9 we see that Kirchhoff's Voltage Law in the presence of voltage sources is, therefore, equivalent to the simple proposition,

$$v - e \text{ is a coboundary.}$$

To obtain a precise formulation of Kirchhoff's Current Law in the presence of voltage sources, let b be any vertex of V. To find the precise analogue of the total current flow into the vertex of the resistive network corresponding to b, we must find all directed branches of which b is a vertex. Such directed branches are obtained from $\delta(\varPhi_b^V)$, the coboundary of the canonical base element $\varPhi_b^V$.

Consider any directed branch (x, y) of K. Recall from the definition of the coboundary operator in 5.6 that the number $\left(\delta(\varPhi_b^V)\right)(x, y)$ is equal to 0 if b is not a vertex of the directed branch (x, y); on the other hand, if b is a vertex of the directed branch (x, y), the number $\left(\delta(\varPhi_b^V)\right)(x, y)$ is equal to 1 or -1, depending upon whether the branch (x, y) is directed towards b or away from b. Thus, to obtain the contribution of the branch (x, y) to the precise analogue of the total current flow into the vertex of the resistive network corresponding to b, we use the product,

$$i(x, y) \left(\delta(\varPhi_b^V)\right)(x, y).$$

The precise analogue of the total current flow into the vertex of the resistive network corresponding to b is the sum,

$$\sum_{(x,y) \in K} i(x, y) \left(\delta(\varPhi_b^V)\right)(x, y),$$

which is equal to the inner product,

$$\langle i, \delta(\Phi_b^V)\rangle.$$

Thus, Kirchhoff's Current Law in the presence of voltage sources requires that

$$\langle i, \delta(\Phi_b^V)\rangle = 0,$$

for each vertex $b \in V$.

The preceding statement can be put in more convenient form. To arrive at such a form consider any coboundary $w \in \mathrm{rng}\ \delta$. Produce $s \in \mathscr{L}(V)$ such that $w = \delta(s)$ and note that $s = \sum_{b \in V} s(b)\ \Phi_b^V$. Hence,

$$\begin{aligned}
\langle i, w\rangle &= \langle i, \delta(s)\rangle \\
&= \langle i, \delta(\sum_{b \in V} s(b)\ \Phi_b^V)\rangle \\
&= \langle i, \sum_{b \in V} s(b)\ \delta(\Phi_b^V)\rangle \\
&= \sum_{b \in V} s(b)\ \langle i, \delta(\Phi_b^V)\rangle.
\end{aligned}$$

Thus, the statement that

$$\langle i, \delta(\Phi_b^V)\rangle = 0$$

for each vertex $b \in V$, implies that

$$\langle i, w\rangle = 0,$$

for each coboundary $w \in \mathrm{rng}\ \delta$.

On the other hand, since $\delta(\Phi_b^V)$ is a coboundary for each $b \in V$, the statement that

$$\langle i, \delta(\Phi_V^b)\rangle = 0$$

for each vertex $b \in V$, is also equivalent to the statement that

$$\langle i, w\rangle = 0$$

for each coboundary $w \in \mathrm{rng}\ \delta$.

But by Theorem 2 of 5.9, this last statement implies that i is a cycle. Finally, reference to Theorem 1 of 5.9 shows that Kirchhoff's Current Law in the presence of voltage sources is, therefore, equivalent to the simple proposition,

$$i \text{ is a cycle.}$$

6.7 Kirchhoff's Laws for Current Sources

Manufacture an actual resistive network from the network (V, S) and the resistance function r, and suppose that the resistive network is inserted into a larger, more complicated circuit in such a way that all energy introduced into the branches of the resistive network is in the form of currents. As in 6.6, the branch currents and branch voltages developed must satisfy, in addition to Ohm's Law, Kirchhoff's Voltage Law and Kirchhoff's Current Law.

Now, in the presence of external currents only, Kirchhoff's Voltage Law, as stated in elementary electrical engineering, demands that the sum of the voltage drops along any closed path of the resistive network must be equal to zero, since no external voltages are introduced along the path. Kirchhoff's Current Law in the presence of external currents only, as stated in elementary electrical engineering, demands that the total branch current flow into each vertex of the resistive network must be equal to the external current flowing into this vertex. In this paragraph we translate these two laws of Kirchhoff into a precise formulation consistent with our abstract setting.

To make matters precise, suppose that the current source chain is the element q of $\mathscr{L}(K)$. As in 6.6 let $i \in \mathscr{L}(K)$ be the current chain induced by q and let $v \in \mathscr{L}(K)$ be the voltage chain induced by q.

For any cycle $u \in$ kernel Δ the precise analogue of the sum of the voltage drops along the closed path of the resistive network corresponding to the cycle u is, as in 6.6, the inner product,

$$\langle v, u \rangle.$$

Hence, Kirchhoff's Voltage Law in the presence of current sources requires that

$$\langle v, u \rangle = 0,$$

for each cycle $u \in$ kernel Δ.

But by the Corollary of 5.11, the preceding statement implies that v is a coboundary. By Theorem 1 of 5.9, we see that Kirchhoff's Voltage Law in the presence of current sources is, therefore, equivalent to the simple proposition,

$$v \text{ is a coboundary.}$$

To obtain a precise formulation of Kirchhoff's Current Law in the presence of current sources we consider any vertex $b \in V$. As in 6.6, the precise analogue of the total branch current flow into the vertex of the resistance network corresponding to b is the inner product

$$\langle i, \delta(\Phi_b^V) \rangle.$$

Similarly, the precise analogue of the total external current flow into the vertex of the resistive network corresponding to b is the inner product

$$\langle q, \delta(\Phi_b^V) \rangle.$$

Thus, Kirchhoff's Current Law in the presence of current sources requires that

$$\langle i, \delta(\Phi_b^V)\rangle = \langle q, \delta(\Phi_b^V)\rangle,$$

for each vertex $b \in V$, which is equivalent to the proposition that

$$\langle i - q, \delta(\Phi_b^V)\rangle = 0,$$

for each vertex $b \in V$.

But as in 6.6, this latter proposition is equivalent to the statement that

$$\langle i - q, w\rangle = 0,$$

for each coboundary $w \in \mathrm{rng}\ \delta$.

But by Theorem 2 of 5.9, this last statement implies that $i - q$ is a cycle. Finally, reference to Theorem 1 of 5.9 shows that Kirchhoff's Current Law in the presence of current sources is, therefore, equivalent to the simple proposition,

$$i - q \text{ is a cycle}.$$

CHAPTER SEVEN

Existence and Uniqueness of Solutions

7.0 Introduction

In Chapter 6 we have repeatedly postponed an investigation of the existence and uniqueness of the current chains and voltage chains describing the performance of a resistive network, but subject to the restrictions imposed by Ohm's Law and Kirchhoff's Current and Voltage Laws. In the first part of this Chapter we finally consider such matters. These results lead naturally into an investigation of network variables, which concludes the Chapter.

7.1 Assumptions of This Chapter

As in Chapter 6 we agree to let R be the field of scalars for our linear algebra.

As in Chapter 5 and Chapter 6 we fix an incidence function f, and agree to let

$$S = \mathrm{dmn}\, f,$$
$$V = \sigma\,\mathrm{dmn}\, f = \sigma S,$$
$$K = \mathrm{rng}\, f.$$

As in 6.2, we fix a resistance function r. Thus, we agree that

$$r \text{ is a function,}$$
$$\mathrm{dmn}\, r = S,$$
$$\mathrm{rng}\, r \subset \{x \mid x \in R \text{ and } x > 0\}.$$

The impedance function Z described in 6.4 will be of use to us. Thus, we agree to fix a function Z such that

$$Z \text{ is the unique linear map of } \mathscr{L}(K) \text{ into } \mathscr{L}(K) \text{ such that}$$
$$\text{for each } (x, y) \in K,$$
$$Z(\Phi^K_{(x,y)}) = r(\{x, y\})\, \Phi^K_{(x,y)}.$$

Two more functions will be of use to us in this Chapter. Recall from the Theorem of 5.11 that

$$\mathscr{L}(K) = \text{kernel } \Delta \oplus \text{rng } \delta.$$

Thus, by 4.8, if $u \in \mathscr{L}(K)$, there exists a unique (v, w) such that

$$v \in \text{kernel } \Delta,$$
$$w \in \text{rng } \delta,$$
$$u = v + w.$$

In this way, two functions L and H can be manufactured, each with domain $\mathscr{L}(K)$, such that for each $u \in \mathscr{L}(K)$,

$$u = L(u) + H(u),$$
$$L(u) \in \text{kernel } \Delta.$$
$$H(u) \in \text{rng } \delta.$$

We agree, throughout this Chapter, to fix L and H such that L and H are these two functions, each with domain $\mathscr{L}(K)$, satisfying the above conditions.

7.2 Linearity of L and H

In this paragraph we show that L and H are both linear maps of $\mathscr{L}(K)$ into $\mathscr{L}(K)$.

Theorem.

(1) L *is a linear map of* $\mathscr{L}(K)$ *into* $\mathscr{L}(K)$.

(2) H *is a linear map of* $\mathscr{L}(K)$ *into* $\mathscr{L}(K)$.

Proof: L is a function and dmn $K = \mathscr{L}(K)$ and rng $L \subset \text{kernel } \Delta \subset \mathscr{L}(K)$. Also, H is a function and dmn $H = \mathscr{L}(K)$ and rng $H \subset \text{rng } \delta \subset \mathscr{L}(K)$. Thus, let $u \in \mathscr{L}(K)$ and let $u' \in \mathscr{L}(K)$ and let $a \in R$. It suffices to thow that

$$L(u + u') = L(u) + L(u'),$$
$$H(u + u') = H(u) + H(u'),$$
$$L(a\,u) = a\,L(u),$$
$$H(a\,u) = a\,H(u).$$

Produce $v \in \text{kernel } \Delta$ and $w \in \text{rng } \delta$ such that $u = v + w$.
Produce $v' \in \text{kernel } \Delta$ and $w' \in \text{rng } \delta$ such that $u' = v' + w'$.

Then, $L(u) = v$ and $L(u') = v'$.
Thus, $L(u) + L(u') = v + v'$.

Also, $H(u) = w$ and $H(u') = w'$.
Thus, $H(u) + H(u') = w + w'$.

But $u + u' = (v + w) + (v' + w') = (v + v') + (w + w')$.
Since kernel Δ is a subspace of $\mathscr{L}(K)$, $v + v' \in$ kernel Δ.
Since rng δ is a subspace of $\mathscr{L}(K)$, $w + w' \in$ rng δ.
Thus, $L(u + u') = v + v' = L(u) + L(u')$.
$H(u + u') = w + w' = H(u) + H(u')$.

Also, $a\,u = a(v + w) = a\,v + a\,w$.
Since kernel Δ is a substance of $\mathscr{L}(K)$, $a\,v \in$ kernel Δ.
Since rng δ is a subspace of $\mathscr{L}(K)$, $a\,w \in$ rng δ.
Thus, $L(a\,u) = a\,v = a\,L(u)$.
$H(a\,u) = a\,w = a\,H(u)$.

7.3 Existence and Uniqueness with Voltage Sources

Throughout this paragraph we consider a voltage source chain e of $\mathscr{L}(K)$. The existence problem asks whether there exist a current chain $i \in \mathscr{L}(K)$ and a voltage chain $v \in \mathscr{L}(K)$ satisfying Ohm's Law,

$$v = Z(i),$$

and such that Kirchhoff's Voltage Law,

$$v - e \text{ is a coboundary},$$

and Kirchhoff's Current Law,

$$i \text{ is a cycle},$$

are also satisfied.

In this paragraph as Theorem 1 we establish the existence of such an $i \in \mathscr{L}(K)$ and $v \in \mathscr{L}(K)$. As Theorem 2 we establish the uniqueness of the current chain $i \in \mathscr{L}(K)$ and the voltage chain $v \in \mathscr{L}(K)$ satisfying the above requirements.

We need two preliminary Lemmas.

Lemma 1. *Let* $s \in \mathscr{L}(K)$. *Suppose that* $\langle Z(s), s \rangle = 0$. *Then,*
$$s = \theta_1.$$

Proof: Note that $s = \sum_{(x,\,y) \in K} s(x, y)\, \Phi^K_{(x,\,y)}$.

Then,

$$0 = \langle Z(s), s \rangle = \langle Z\big(\sum_{(x,\, y) \in K} s(x, y)\, \Phi^K_{(x,\, y)} \big), \sum_{(a,\, b) \in K} s(a, b)\, \Phi^K_{(a,\, b)} \rangle$$

$$= \langle \sum_{(x,\, y) \in K} s(x, y)\, Z(\Phi^K_{(x,\, y)}), \sum_{(a,\, b) \in K} s(a, b)\, \Phi^K_{(a,\, b)} \rangle$$

$$= \langle \sum_{(x,\, y) \in K} s(x, y)\, r(\{x, y\})\, \Phi^K_{(x,\, y)}, \sum_{(a,\, b) \in K} s(a, b)\, \Phi^K_{(a,\, b)} \rangle$$

$$= \sum_{(x,\, y) \in K} \sum_{(a,\, b) \in K} s(x, y)\, r(\{x, y\})\, s(a, b)\, \langle \Phi^K_{(x,\, v)}, \Phi^K_{(a,\, b)} \rangle$$

$$= \sum_{(x,\, y) \in K} r(\{x, y\})\, (s(x, y))^2.$$

Thus, since $r(\{x, y\}) > 0$ for each $(x, y) \in K$, we conclude that $s(x, y) = 0$ for each $(x, y) \in K$, implying that $s = \theta_1$.

Lemma 2.

(1) $(L \circ Z) |\text{kernel } \Delta$ *is a linear map of* kernel Δ *into* kernel Δ.

(2) $(L \circ Z) |\text{kernel } \Delta$ *is univalent.*

(3)   *inv* $((L \circ Z) |\text{kernel } \Delta)$ *is a linear map of* kernel Δ *into* kernel Δ.

Proof of (1): We know that Z is a linear map of $\mathscr{L}(K)$ into $\mathscr{L}(K)$. By the Theorem of 7.2, L is a linear map of $\mathscr{L}(K)$ into $\mathscr{L}(K)$. From elementary linear algebra, the composition of two linear maps is a linear map, and thus, $L \circ Z$ a linear map of $\mathscr{L}(K)$ into $\mathscr{L}(K)$.

But rng $(L \circ Z) \subset$ rng $L \subset$ kernel Δ.

Also, kernel Δ is a subspace of $\mathscr{L}(K)$.

Thus, $L \circ Z$ is a linear map of $\mathscr{L}(K)$ into kernel Δ.

But kernel $\Delta \subset \mathscr{L}(K) = \text{dmn } Z = \text{dmn } (L \circ Z)$.

Hence, dmn $((L \circ Z) |\text{kernel } \Delta) = $ kernel Δ.

Finally, elementary linear algebra assures us that the restriction of a linear map to a subspace of its domain is also a linear map, and thus, $(L \circ Z) |\text{kernel } \Delta$ is a linear map of kernel Δ into kernel Δ.

Proof of (2): In view of (1), elementary linear algebra assures us that it suffices to let $s \in$ kernel Δ such that $((L \circ Z) |\text{kernel } \Delta) (s) = \theta_1$, and show that $s = \theta_1$.

Thus, let $s \in$ kernel Δ.

Suppose that $((L \circ Z) |\text{kernel } \Delta) (s) = \theta_1$.

Then, $(L \circ Z)(s) = \theta_1$.

Consider $Z(s) \in \mathscr{L}(K)$.

Produce $v \in$ kernel Δ and $w \in$ rng δ such that $Z(s) = v + w$. Then, $L(Z(s)) = v$.

But we know that $\theta_1 = (L \circ Z)(s) = L(Z(s))$.

Thus, $v = \theta_1$, implying that $Z(s) = w \in$ rng δ.

By Theorem 1 of 5.9, $\langle s, Z(s) \rangle = 0$.

By Lemma 1 of this paragraph, $s = \theta_1$.

Proof of (3): From elementary linear algebra we know that if a univalent function is a linear map of a finite dimensional linear space into itself, then the range of this function must equal its domain. Thus, by (2)

$$\text{rng} \left((L \circ Z)|\text{kernel } \Delta\right) = \text{dmn} \left((L \circ Z)|\text{kernel } \Delta\right) = \text{kernel } \Delta.$$

Hence,

$$\text{dmn} \left(\text{inv}((L \circ Z)|\text{kernel } \Delta)\right) = \text{rng} \left((L \circ Z)|\text{kernel } \Delta\right) = \text{kernel } \Delta.$$

Also,

$$\text{rng} \left(\text{inv}((L \circ Z)|\text{kernel}\Delta)\right) = \text{dmn} \left((L \circ Z)|\text{kernel } \Delta\right) = \text{kernel } \Delta.$$

Finally, elementary linear algebra assures that the inverse of a univalent linear map is also a linear map, and thus, inv $((L \circ Z)|\text{kernel } \Delta)$ is a linear map of kernel Δ into kernel Δ.

In the following Theorem 1, as previously advertised, we establish the existence of an $i \in \text{kernel } \Delta$ and $v \in \mathscr{L}(K)$ such that $v = Z(i)$ and $v - e \in \text{rng } \delta$. In particular, we show that if we let

$$i = \left(\text{inv} \left((L \circ Z)|\text{kernel } \Delta\right)\right) (L(e)),$$

and let

$$v = Z(i),$$

the required conditions are satisfied.

Theorem 1.

(1) $\left(\text{inv}((L \circ Z)|\text{kernel } \Delta)\right) (L(e)) \in \text{kernel } \Delta$.

(2) $Z\left(\left(\text{inv}((L \circ Z)|\text{kernel } \Delta)\right) (L(e))\right) - e \in \text{rng } \delta$.

Proof of (1): By the definition of L in 7.2 and (3) of Lemma 2,

$$L(e) \in \text{kernel } \Delta = \text{dmn} \left(\text{inv} \left((L \circ Z)|\text{kernel } \Delta\right)\right).$$

Thus, by (3) of Lemma 2,

$$\left(\text{inv} \left((L \circ Z)|\text{kernel } \Delta\right)\right) (L(e)) \in \text{rng} \left(\text{inv} \left((L \circ Z)|\text{kernel } \Delta\right)\right) \subset \text{kernel } \Delta.$$

Proof of (2): Let

$$v = Z\left(\left(\text{inv} \left((L \circ Z)|\text{kernel } \Delta\right)\right) (L(e))\right).$$

We must show that $v - e \in \text{rng } \delta$.

But by the definition of H and L,

$$v - e = L(v - e) + H(v - e).$$

Since $H(v - e) \in \text{rng } \delta$, it suffices to show that $L(v - e) = \theta_1$.

But by the linearity of L,

$$
\begin{aligned}
L(v - e) &= L(v) - L(e) \\
&= L\Big(Z\big((\mathrm{inv}((L \circ Z)|\mathrm{kernel}\ \varDelta))\,(L(e))\big)\Big) - L(e) \\
&= \big((L \circ Z) \circ (\mathrm{inv}((L \circ Z)|\mathrm{kernel}\ \varDelta))\big)\,(L(e)) - L(e) \\
&= \Big(((L \circ Z) \circ (\mathrm{inv}(L \circ Z)))|\mathrm{kernel}\ \varDelta\Big)\,(L(e)) - L(e) \\
&= ((L \circ Z) \circ (\mathrm{inv}(L \circ Z)))\,(L(e)) - L(e) \\
&= L(e) - L(e) \\
&= \theta_1.
\end{aligned}
$$

We have shown that if we let

$$
i = \big(\mathrm{inv}((L \circ Z)|\mathrm{kernel}\ \varDelta)\big)\,(L(e)),
$$

and let

$$
v = Z(i),
$$

then $i \in \mathrm{kernel}\ \varDelta$ and $v - e \in \mathrm{rng}\ \delta$.

To establish the uniqueness of such an i and v in Theorem 2 we show that if $j \in \mathrm{kernel}\ \varDelta$ and $w \in \mathscr{L}(K)$ such that $w = Z(j)$ and $w - e \in \mathrm{rng}\ \delta$, then $j = i$ and $v = w$.

Clearly, in such an argument it suffices to show only that $j = i$, for then we would have

$$
w = Z(j) = Z(i) = v.
$$

Theorem 2. *Let* $j \in \mathrm{kernel}\ \varDelta$. *Suppose that* $Z(j) - e \in \mathrm{rng}\ \delta$. *Then,*

$$
j = \big(\mathrm{inv}((L \circ Z)|\mathrm{kernel}\ \varDelta)\big)\,(L(e)).
$$

Proof: Let

$$
i = \big(\mathrm{inv}((L \circ Z)|\mathrm{kernel}\ \varDelta)\big)\,(L(e)).
$$

By (2) of Theorem 1 of this paragraph, $Z(i) - e \in \mathrm{rng}\ \delta$.
Produce $s \in \mathrm{rng}\ \delta$ such that $Z(i) - e = s$.
Then, $Z(i) = s + e$.
But also, $Z(j) - e \in \mathrm{rng}\ \delta$.
Produce $t \in \mathrm{rng}\ \delta$ such that $Z(j) - e = t$.
Then, $Z(j) = t + e$.
Hence, $Z(j) - Z(i) = (t + e) - (s + e) = t - s$.
Since $\mathrm{rng}\ \delta$ is a subspace of $\mathscr{L}(K)$, $Z(j) - Z(i) \in \mathrm{rng}\ \delta$.
By the linearity of Z, $Z(j - i) = Z(j) - Z(i) \in \mathrm{rng}\ \delta$.
But $j \in \mathrm{kernel}\ \varDelta$.
By (1) of Theorem 1 of this paragraph, $i \in \mathrm{kernel}\ \varDelta$.
Since $\mathrm{kernel}\ \varDelta$ is a subspace of $\mathscr{L}(K)$, $j - i \in \mathrm{kernel}\ \varDelta$.
Thus, by Theorem 1 of 5.9, $\langle j - i, Z(j - i)\rangle = 0$.
By Lemma 1 of this paragraph, $j - i = \theta_1$, implying that $j = i$.

7.4 Existence and Uniqueness with Current Sources

Throughout this paragraph we consider a current source chain q of $\mathscr{L}(K)$. The existence problem asks whether there exist a current chain $i \in \mathscr{L}(K)$ and a voltage chain $v \in \mathscr{L}(K)$ satisfying Ohm's Law,

$$v = Z(i),$$

and such that Kirchhoff's Voltage Law,

$$v \text{ is a coboundary},$$

and Kirchhoff's Current Law,

$$i - q \text{ is a cycle},$$

are also satisfied.

In this paragraph as Theorem 1 we establish the existence of such an $i \in \mathscr{L}(K)$ and $v \in \mathscr{L}(K)$. As Theorem 2 we establish the uniqueness of the current chain $i \in \mathscr{L}(K)$ and the voltage chain $v \in \mathscr{L}(K)$ satisfying the above requirements.

We need three preliminary Lemmas.

Lemma 1. *Let Y be the unique linear map of $\mathscr{L}(K)$ into $\mathscr{L}(K)$ such that for each $(x, y) \in K$,*

$$Y(\Phi^K_{(x,y)}) = (r(\{x, y\}))^{-1} \, \Phi^K_{(x,y)}.$$

Then,

(1) $Y = \operatorname{inv} Z$.

(2)   *Z is univalent.*

Proof: Since Y is a function, (2) is trivially implied by (1). Thus, we only prove (1).

Note that $\operatorname{dmn} Z = \operatorname{dmn} Y = \mathscr{L}(K)$. Thus, to show that

$$Y \subset \operatorname{inv} Z \text{ and } \operatorname{inv} Z \subset Y,$$

it suffices to show that for each $s \in \mathscr{L}(K)$,

$$Y(Z(s)) = Z(Y(s)) = s.$$

Let $s \in \mathscr{L}(K)$.

But,

$$
\begin{aligned}
Y(Z(s)) &= Y\Big(Z\Big(\sum_{(x,y) \in K} s(x, y)\, \Phi^K_{(x,y)}\Big)\Big) \\
&= Y\Big(\sum_{(x,y) \in K} s(x, y)\, Z(\Phi^K_{(x,y)})\Big) \\
&= Y\Big(\sum_{(x,y) \in K} s(x, y)\, r(\{x, y\})\, \Phi^K_{(x,y)}\Big)
\end{aligned}
$$

$$= \sum_{(x,\,y)\,\in\,K} s(x,\,y)\, r(\{x,\,y\})\, Y(\Phi^K_{(x,\,y)})$$

$$= \sum_{(x,\,y)\,\in\,K} s(x,\,y)\, r(\{x,\,y\})\, (r(\{x,\,y\}))^{-1}\, \Phi^K_{(x,\,y)}$$

$$= \sum_{(x,\,y)\,\in\,K} s(x,\,y)\, \Phi^K_{(x,\,y)}$$

$$= s.$$

Similarly, it is easy to show that $Z(Y(s)) = s$.

Lemma 2. *Let $Y = \mathrm{inv}\, Z$. Let $s \in \mathscr{L}(K)$. Suppose that $\langle Y(s), s \rangle = 0$. Then,*

$$s = \theta_1.$$

Proof: Since $Y = \mathrm{inv}\, Z$, $Z(Y(s)) = s$.
Thus, $\langle Y(s), Z(Y(s)) \rangle = 0$.
By Lemma 1 of 7.3, $Y(s) = \theta_1$.
Then, $s = Z(Y(s)) = Z(\theta_1)$.
But since Z is a linear map of $\mathscr{L}(K)$ into $\mathscr{L}(K)$, $Z(\theta_1) = \theta_1$.

Lemma 3. *Let $Y = \mathrm{inv}\, Z$.*
Then,

    (1) $(H \circ Y)|\,\mathrm{rng}\,\delta$ is a linear map of $\mathrm{rng}\,\delta$ into $\mathrm{rng}\,\delta$.

    (2) $(H \circ Y)|\,\mathrm{rng}\,\delta$ is univalent.

    (3) $\mathrm{inv}\,((H \circ Y)|\,\mathrm{rng}\,\delta)$ is a linear map of $\mathrm{rng}\,\delta$ into $\mathrm{rng}\,\delta$.

Proof of (1): By Lemma 1 of this paragraph, Y is a linear map of $\mathscr{L}(K)$ into $\mathscr{L}(K)$.
From the Theorem of 7.2, H is a linear map of $\mathscr{L}(K)$ into $\mathscr{L}(K)$.
Thus, the composition, $H \circ Y$, is a linear map of $\mathscr{L}(K)$ into $\mathscr{L}(K)$.
    But $\mathrm{rng}\,(H \circ Y) \subset \mathrm{rng}\,H \subset \mathrm{rng}\,\delta$.
Also, $\mathrm{rng}\,\delta$ is a subspace of $\mathscr{L}(K)$.
Thus, $H \circ Y$ is a linear map of $\mathscr{L}(K)$ into $\mathrm{rng}\,\delta$.
    But $\mathrm{rng}\,\delta \subset \mathscr{L}(K) = \mathrm{dmn}\,Y = \mathrm{dmn}\,(H \circ Y)$.
Thus, $\mathrm{dmn}\,((H \circ Y)|\,\mathrm{rng}\,\delta) = \mathrm{rng}\,\delta$.
    Finally, $(H \circ Y)|\,\mathrm{rng}\,\delta$, the restriction of a linear map, is a linear map of $\mathrm{rng}\,\delta$ into $\mathrm{rng}\,\delta$.

Proof of (2): Let $s \in \mathrm{rng}\,\delta$ such that $(H \circ Y)(s) = \theta_1$.
It suffices to show that $s = \theta_1$.
    Consider $Y(s) \in \mathscr{L}(K)$.
Produce $v \in \mathrm{kernel}\,\Delta$ and $w \in \mathrm{rng}\,\delta$ such that $Y(s) = v + w$. Then, $H(Y(s)) = w$.
    But we know that $\theta_1 = (H \circ Y)(s) = H(Y(s))$.
Thus, $w = \theta_1$, implying that $Y(s) = v \in \mathrm{kernel}\,\Delta$.
But by Theorem 1 of 5.9, $\langle Y(s), s \rangle = 0$.
By Lemma 2 of this paragraph, $s = \theta_1$.

Proof of (3): From elementary linear algebra, using (2),

$$\operatorname{rng}\big((H{\circ}Y)|\operatorname{rng}\delta\big) = \operatorname{dmn}\big((H{\circ}Y)|\operatorname{rng}\delta\big) = \operatorname{rng}\delta.$$

Hence,

$$\operatorname{dmn}\big(\operatorname{inv}((H{\circ}Y)|\operatorname{rng}\delta)\big) = \operatorname{rng}\big((H{\circ}Y)|\operatorname{rng}\delta\big) = \operatorname{rng}\delta.$$

Also,

$$\operatorname{rng}\big(\operatorname{inv}((H{\circ}Y)|\operatorname{rng}\delta)\big) = \operatorname{dmn}\big((H{\circ}Y)|\operatorname{rng}\delta\big) = \operatorname{rng}\delta.$$

Since the inverse of a univalent linear map is also a linear map, $\operatorname{inv}\big((H{\circ}Y)|\operatorname{rng}\delta\big)$ is a linear map of $\operatorname{rng}\delta$ into $\operatorname{rng}\delta$.

In the following Theorem 1, as previously advertised, we establish the existence of an $i \in \mathscr{L}(K)$ and $v \in \operatorname{rng}\delta$ such that $v = Z(i)$ and $i - q \in \operatorname{kernel}\Delta$. In particular, we let $Y = \operatorname{inv}Z$, and then show that if we let

$$v = \big(\operatorname{inv}((H{\circ}Y)|\operatorname{rng}\delta)\big)(H(q)),$$

and let

$$i = Y(v),$$

the required conditions are satisfied.

Theorem 1. *Let* $Y = \operatorname{inv}Z$.
Then,

    (1) $\big(\operatorname{inv}((H{\circ}Y)|\operatorname{rng}\delta)\big)(H(q)) \in \operatorname{rng}\delta$.

    (2) $Y\big((\operatorname{inv}((H{\circ}Y)|\operatorname{rng}\delta))(H(q))\big) - q \in \operatorname{kernel}\Delta$.

Proof of (1): From the definition of H in 7.2 and (3) of Lemma 3,

$$H(q) \in \operatorname{rng}\delta = \operatorname{dmn}\big(\operatorname{inv}((H{\circ}Y)|\operatorname{rng}\delta)\big).$$

Thus, by (3) of Lemma 3,

$$\big(\operatorname{inv}((H{\circ}Y)|\operatorname{rng}\delta)\big)(H(q)) \in \operatorname{rng}\big(\operatorname{inv}((H{\circ}Y)|\operatorname{rng}\delta)\big) \subset \operatorname{rng}\delta.$$

Proof of (2): Let

$$i = Y\big((\operatorname{inv}((H{\circ}Y)|\operatorname{rng}\delta))(H(q))\big).$$

We must show that $i - q \in \operatorname{kernel}\Delta$.

But by the definition of L and H,

$$i - q = L(i - q) + H(i - q).$$

Since $L(i - q) \in \operatorname{kernel}\Delta$, it suffices to show that $H(i - q) = \theta_1$. But by the linearity of H,

$$H(i - q) = H(i) - H(q)$$
$$= H\Big(Y\big((\text{inv}\,((H \circ Y)|\text{rng }\delta))\,(H(q))\big)\Big) - H(q)$$
$$= \big((H \circ Y) \circ (\text{inv}\,((H \circ Y)|\text{rng }\delta))\big)\,(H(q)) - H(q)$$
$$= \big(((H \circ Y) \circ (\text{inv}\,(H \circ Y)))|\text{rng }\delta\big)\,(H(q)) - H(q)$$
$$= \big((H \circ Y) \circ (\text{inv}\,(H \circ Y))\big)\,(H(q)) - H(q)$$
$$= H(q) - H(q)$$
$$= \theta_1.$$

We have shown that if we let $Y = \text{inv}\,Z$ and let

$$v = \big(\text{inv}\,((H \circ Y)|\text{rng }\delta)\big)\,(H(q)),$$

and let

$$i = Y(v),$$

then $v = Z(i)$ and $v \in \text{rng }\delta$ and $i - q \in \text{kernel }\Delta$.

To establish the uniqueness of such an i and v in Theorem 2 we show that if $w \in \text{rng }\delta$ and $j \in \mathscr{L}(K)$ such that $w = Z(j)$ and $j - q \in \text{kernel }\Delta$, then $j = i$ and $w = v$.

Clearly, in such an argument it suffices to show only that $w = v$, for then we would have

$$j = Y(w) = Y(v) = i.$$

Theorem 2. *Let* $Y = \text{inv}\,Z$. *Let* $w \in \text{rng }\delta$. *Suppose that* $Y(w) - q \in \text{kernel }\Delta$.
Then,

$$w = \big(\text{inv}\,((H \circ Y)|\text{rng }\delta)\big)\,(H(q)).$$

Proof: Let

$$v = \big(\text{inv}\,((H \circ Y)|\text{rng }\delta)\big)\,(H(q)).$$

By (2) of Theorem 1 of this paragraph, $Y(v) - q \in \text{kernel }\Delta$. Produce $s \in \text{kernel }\Delta$ such that $Y(v) - q = s$.
Then, $Y(v) = s + q$.

But also, $Y(w) - q \in \text{kernel }\Delta$.
Produce $t \in \text{kernel }\Delta$ such that $Y(w) - q = t$.
Then, $Y(w) = t + q$.

Hence, $Y(w) - Y(v) = (t + q) - (s + q) = t - s$.
Since kernel Δ is a subspace of $\mathscr{L}(K)$, $Y(w) - Y(v) \in \text{kernel }\Delta$.
By the linearity of Y, $Y(w - v) = Y(w) - Y(v) \in \text{kernel }\Delta$.

But $w \in \text{rng }\delta$.

By (1) of Theorem 1 of this paragraph, $v \in \mathrm{rng}\ \delta$.
Since $\mathrm{rng}\ \delta$ is a subspace of $\mathscr{L}(K)$, $w - v \in \mathrm{rng}\ \delta$.

By Theorem 1 of 5.9, $\langle Y(w - v), w - v \rangle = 0$.
By Lemma 2 of this paragraph, $w - v = \theta_1$, implying that $w = v$.

7.5 Current Variables

Suppose that either a current source chain or a voltage source chain is applied to our network (V, S) inducing a current chain $i \in \mathscr{L}(K)$ and a voltage chain $v \in \mathscr{L}(K)$ satisfying Ohm's Law and Kirchhoff's Voltage Law and Kirchhoff's Current Law. To describe the performance of the network we must know the numbers $i(x, y)$ and $v(x, y)$ for each directed branch (x, y) of K.

The reader may think that these problems have been solved in Paragraphs 7.3 and 7.4. For example, in 7.3, in the case of a voltage source chain e of $\mathscr{L}(K)$, we are assured that

$$i(x, y) = \Big((\mathrm{inv}\,((L \circ Z)|\mathrm{kernel}\ \varDelta))\,(L(e))\Big)(x, y),$$

and

$$v(x, y) = (Z(i))(x, y),$$

for each $(x, y) \in K$, while in 7.4, in the case of a current source chain q of $\mathscr{L}(K)$, we are assured that with $Y = \mathrm{inv}\ Z$,

$$v(x, y) = \Big((\mathrm{inv}\,((H \circ Y)|\mathrm{rng}\ \delta))\,(H(q))\Big)(x, y),$$

and

$$i(x, y) = (Y(v))(x, y),$$

for each $(x, y) \in K$.

However, the above formulas, while useful for establishing existence and uniqueness theorems, are not of much practical use in providing us with the numbers $i(x, y)$ and $v(x, y)$ for each directed branch $(x, y) \in K$. These formulas fail to be useful because they depend upon the functions L and H, and although L and H certainly exist, they are not readily available for practical use in describing the performance of the network.

Thus, we must seek other more practical, explicit formulas for the numbers $i(x, y)$ and $v(x, y)$ for each directed branch $(x, y) \in K$. In the following Chapter Eight we exhibit such explicit formulas which can be used to calculate the numbers $i(x, y)$ and $v(x, y)$ for each directed branch $(x, y) \in K$ in any arbitrary network of any complexity.

In the remainder of this Chapter we describe and make precise certain techniques which the electrical engineer uses to compute the numbers $i(x, y)$ and $v(x, y)$ for each directed branch $(x, y) \in K$ in networks with a relatively simple structure. In this Paragraph 7.5 we discuss these techniques with reference to the current chain $i \in \mathscr{L}(K)$, while in the following Paragraph 7.6 we discuss these techniques with reference to the voltage chain $v \in \mathscr{L}(K)$.

Thus, we now assume that an electrical engineer is given a reasonably simple resistive network energize dexclusively by voltage sources. He is required to find the branch current $i(x, y)$ for each directed branch (x, y) of the network.

When an electrical engineer subjects a resonably simple resistive network, energized exclusively by voltage sources, to analysis, he usually does not try to compute directly the value of the curent chain $i(x, y)$ for each directed branch (x, y) of the network. Instead, he introduces some auxiliary current variables called loop current variables. In particular, he selects a set of loops of the network, and then noting that no meaning has yet been assigned to the term "loop current", he assigns a loop current to each of those loops selected. We describe the process in greater detail.

The loops selected for the auxiliary loop current variables must be selected with some care. The engineer must select a set of loops with the property that each loop of the set contains a designated branch which is not contained in any other loop of the set. In each such loop, he considers the value of the branch current of this designated branch, and agrees that the loop current of that loop will be set equal to the value of this designated branch current. In this way a set of auxiliary loop current variables is manufactured.

A set of such loops with the property that each loop of the set contains a designated branch which is not contained in any other loop of the set is usually called an independent set of loops. The word "independent" is used to suggest that no member of the set of loops can be manufactured from the remaining loops of the set by some vague process involving the combining of branches from the remaining loops of the set.

The engineer must also select exactly the right number of auxiliary loop current variables to be used in the analysis. Since each independent loop gives rise to one auxiliary loop current variable, he must specify the number of independent loops to be selected. This magic number is usually specified as the maximal number of such independent loops which can be produced from the network.

Observe that we have selected certain loops of the network in our set of independent loops. There are, of course, many other loops of the network which are not members of our set of independent loops. Now, if a branch of the network is such that it is contained in no loop of the network, then the engineer knows that no current will flow in that branch and the value of the branch current is zero. Thus, the engineer is only interested in the values of the branch currents of those branches which are contained in some loop of the network. A maximal set of independent loops can be used to determine the values of the branch currents of such branches.

In particular, a maximal set of independent loops has another property called completeness. By completeness, we mean that each branch current of a branch contained in some loop of the network can be written as a sum of the auxiliary loop current variables. More precisely, each such branch current is the sum of the auxiliary loop current variables corresponding to those loops of the independent set of loops which contain the branch in question. Thus, to determine all branch currents it suffices to find the value of each of the auxiliary loop current variables.

Instead of seeking a maximal set of independent loops, the engineer could look for a set of independent loops which is also complete. Such a set of loops is automatically a maximal set of independent loops. Alternatively, he could look for a minimal set of complete loops. Such a set would automatically be an independent set of loops. We assume that by any of these processes the engineer has established an appropriate set of loops and corresponding auxiliary loop current variables.

By applying Kirchhoff's Voltage Law and Ohm's Law to each loop of the auxiliary loop current variables, and expressing each branch current involved in terms of the auxiliary loop current variables, the engineer obtains a system of linear equations in these auxiliary loop current variables. There are precisely as many linear equations in this system as there are auxiliary loop current variables. This system of linear equations is then solved to yield the values of the auxiliary loop current variables. The branch current can then be obtained from the values of the auxiliary loop current variables as described above.

Although the selection of an appropriate set of auxiliary loop current variables appears to be a difficult job, it is a remarkable fact that the practicing engineer can usually apply this process to analyze a reasonably simple network and obtain the values of the branch currents. For such simple networks the engineer instinctively knows how to select an appropriate maximal independent, or minimal complete set of loops for the auxiliary loop current variables. It is our intention here to show that this popular process involving the selection of auxiliary loop current variables is soundly based upon precise principles which we can derive from our abstract formulation.

In our abstract formulation, the precise analogue of a loop is a cycle. An independent set of loops translates into a linearly independent set of cycles, while a complete set of loops translates into a generating set of cycles. Thus, a set of loops which is independent and complete has as its precise analogue a linearly independent generating set of cycles, which is, of course, a base of the cycle space, kernel Δ. Alternatively, a maximal independent set of loops corresponds to a maximal linearly independent set of cycles, which is a base, while a minimal complete set of loops corresponds to a minimal generating set of cycles, which is a base. Thus, the precise analogue of a set of loops appropriate for the auxiliary loop current variables is merely a base of the cycle space, kernel Δ.

The number of auxiliary loop current variables is equal to the number of elements of any base of the cycle space, kernel Δ, and this number is equal to the dimension of the cycle space, kernel Δ. We remind the reader that we have investigated the dimension of kernel Δ in Chapter 5. In particular, we showed in the Corollary of 5.14 that if (V, S) is a connected network,

$$\dim (\text{kernel } \Delta) = \dim (\mathscr{L}(K)) - \dim (\mathscr{L}(V)) + 1.$$

Since the dimension of $\mathscr{L}(K)$ is equal to the number of elements of K which is equal to the number of elements of S, while the dimension of $\mathscr{L}(V)$ is equal to the number of elements of V, we have, when (V, S) is a connected network,

$$\dim (\text{kernel } \Delta) = pS - pV + 1.$$

Thus, if (V, S) is a connected network, the number of auxiliary loop current variables is equal to one more than the number of branches diminished by the number of vertices.

Since our unknown current chain i is a cycle, we can express i as a linear combination of the elements of any base of the cycle space, kernel Δ. This fact is the precise analogue of our earlier assertion that each branch current can be written as a linear combination of the auxiliary loop current variables.

Now, there are many different bases for the cycle space, kernel Δ. Thus, the engineer could select many different sets of auxiliary loop current variables to be used in his analysis. We shall describe one process which the engineer uses to produce an appropriate set of auxiliary loop current variables. Then we shall show that a precise analogue of this process can be used to construct a base of the cycle space, kernel Δ. In fact, the particular base of the cycle space, kernel Δ, that we construct, will be of use to us in our work in the following Chapter Eight.

For the remainder of this Paragraph we assume that (V, S) is a connected network. We exhibit first a process used by the engineer to manufacture an appropriate set of auxiliary loop current variables.

Let T be a tree in (V, S) and suppose that each vertex of V is a vertex of T. Now, we know that each branch of T is a branch of S. If the branches of T comprise the complete set of branches S, then $T = S$, the network (V, S) has no loops, and is of little practical interest to the engineer since no current will flow. Thus, we assume that $T \neq S$, and consider a branch $\{x, y\}$ of S such that $\{x, y\}$ is not a branch of T. By the Theorem of 2.12, there exists a unique loop W such that $\{x, y\}$ is a branch of W and all other branches of W are branches of T.

For each branch $\{x, y\}$ of S which is not a branch of T manufacture such a unique loop W. Clearly, all such manufactured loops will be distinct. There will be exactly as many such manufactured loops as there are branches of S which are not branches of T. Thus, the number of such manufactured loops is equal to $pS - pT$, the number of elements of S diminished by the number of elements of T.

But we know from Theorem 1 of 2.10 that the number of branches of T is one less than the number of vertices of T. Our assumption that each vertex of V is a vertex of T means that the vertices of T are precisely the vertices of V. Thus, the number of branches of T is one less than the number of vertices of V, and hence, the number of such manufactured loops is equal to

$$pS - pV + 1,$$

which is the magic number of loops needed by the engineer to establish an appropriate set of auxiliary loop current variables. Furthermore, this set of $pS - pV + 1$ manufactured loops is independent, in the sense demanded by the engineer in order to establish a set of auxilary loop current variables. Thus, by assigning a current variable to each of these $pS - pV + 1$ manufactured loops the engineer obtains an appropriate set of auxiliary loop current variables.

It is worthwhile to note that the process described above depends upon the tree T used. Thus, for each such tree T, an appropriate set of auxiliary loop current variables can be manufactured. Therefore, there are at least as many different choices of auxiliary loop current variables as there are trees whose vertices comprise all the vertices of V. However, it turns out that there are more many sets of auxiliary loop current variables than those which can be manufactured from trees by the process described above.

We now exhibit a precise analogue of the preceding process, to manufacture a base of the cycle space, kernel Δ.

Let T be a tree in (V, S) whose vertices comprise all the vertices of V. Thus, $\sigma T = V$.

Now, if $T = S$, then by Theorem 1 of 2.10,

$$\dim (\text{kernel } \varDelta) = pS - pV + 1 = pT - p\sigma T + 1 = 0,$$

and the cycle space, kernel $\varDelta$, is trivial and of no interest. Thus, we assume that $T \neq S$.

As before, for each branch $\{x, y\} \in S - T$, we apply the Theorem of 2.12 and consider the unique loop which includes $\{x, y\}$ as a branch, and whose other branches are branches of T. Throughout the remainder of this book we shall have need for this unique loop, and thus, we adopt $\lambda(T, \{x, y\})$ as a symbol for this loop. Formally,

$\lambda(T, \{x, y\})$ is such that

(1) $\lambda(T, \{x, y\})$ is a loop.

(2) $\{x, y\} \in \lambda(T, \{x, y\}) \subset T \cup \{\{x, y\}\}$.

Consider some branch $\{x, y\} \in S - T$ and the loop $\lambda(T, \{x, y\})$. From this loop we want to manufacture a cycle which will be an element of our base of the cycle space, kernel $\varDelta$. This cycle to be manufactured from $\lambda(T, \{x, y\})$ must be an element of $\mathscr{L}(K)$, and thus, it must be a function whose domain is equal to K and whose range consists of real numbers. As a first step in the specification of this function it seems reasonable for the function to assign the number 0 to all directed branches (s, t) of K such that $\{s, t\}$ is not an element of the loop $\lambda(T, \{x, y\})$.

Thus, we must decide how this function will assign numbers to those directed branches (s, t) of K such that $\{s, t\}$ is an element of the loop $\lambda(T, \{x, y\})$. Now, each directed branch (s, t) already has a direction assigned to it by the incidence function f, since $(s, t) = f(\{s, t\})$. By assigning a positive number to the directed branch (s, t) we retain its direction; by assigning a negative number to the directed branch (s, t) we reverse its direction. To obtain a cycle from an assignment of numbers to those directed branches (s, t) of K such that $\{s, t\}$ is an element of $\lambda(T, \{x, y\})$, it suffices to have the same magnitude for all of the numbers assigned to these directed branches, and to arrange the algebraic signs of the numbers assigned to these directed branches such that after the assignment of the numbers to the directed branches, all branches will have the same direction during a trip around the loop.

To simplify matters we can assign a number of magnitude 1 to each directed branch (s, t) of K such that $\{s, t\}$ is an element of the loop $\lambda(T, \{x, y\})$. A proper selection of algebraic signs of the numbers to be assigned to each such directed branch (s, t) will result in either all branches pointing in a clockwise direction or all branches pointing in a counter-clockwise direction during a trip around the loop. Since we have this choice of direction we shall simplify matters by selecting the algebraic signs of the numbers assigned to the directed branches in such a way that all branches have the direction of the directed branch $f(\{x, y\})$.

Throughout the remainder of the book we shall adopt the symbol $c(T, \{x, y\})$ for this cycle to be manufactured from the loop $\lambda(T, \{x, y\})$. Thus, we must now distill the prose of the preceding paragraphs to obtain a formal definition of $c(T, \{x, y\})$, and then show as a Theorem than $c(T, \{x, y\})$ is indeed a cycle.

We have specified above in loose prose some properties which the function $c(T, \{x, y\})$ must have. However, at this point we do not know whether there even exists a function satisfying the requirements we have set down. Thus, as our first step in the definition of $c(T, \{x, y\})$ we must translate into precise mathematical terms the crucial properties which the function must have, and then show that there does indeed exist a function with these properties. This is accomplished in the following Lemma.

Lemma. *Let M be a loop in (V, S). Let $\{x, y\} \in M$. Then,*

for some u,

(1)   *u is a function.*

(2) $\mathrm{dmn}\ u = \mathrm{rng}\ (f|M)$.

(3) $\mathrm{rng}\ u \subset \{1, -1\}$.

(4) $u(f(\{x, y\})) = 1$.

(5) $(s, t) \in \mathrm{dmn}\ u$ *and* $(t, m) \in \mathrm{dmn}\ u$
     implies
     $u(s, t) = u(t, m)$.

(6) $(s, t) \in \mathrm{dmn}\ u$ *and* $(m, t) \in \mathrm{dmn}\ u$ *and* $(s, t) \neq (m, t)$
     implies
     $u(s, t) = -u(m, t)$.

(7) $(t, s) \in \mathrm{dmn}\ u$ *and* $(t, m) \in \mathrm{dmn}\ u$ *and* $(t, s) \neq (t, m)$
     implies
     $u(t, s) = -u(t, m)$.

Proof: We lose no generality by assuming that $f(\{x, y\}) = (y, x)$. By Theorem 1 and Theorem 2 of 2.5, produce g such that

$$g \text{ is a proper path from } x \text{ to } y \text{ in } M - \{\{x, y\}\},$$
$$\mathrm{rng}\ g = M - \{\{x, y\}\},$$
$$\sigma\mathrm{rng}\ g = \sigma M.$$

Let $n \in \omega$ such that $\mathrm{dmn}\ g = [n]$.

By the Theorem of 1.15, establish a univalent sequence b with $\mathrm{dmn}\ b = [n + 1]$ such that

$$b_1 = x,$$
$$b_{n+1} = y,$$
$$\{b_j, b_{j+1}\} = g_j, \quad \text{for each } j \in [n].$$

Next, we construct the function u with dmn u = rng $(f|M)$ as follows:

First, let $u(y, x) = 1$.

Next, let $j \in [n]$ and examine $f(\{b_j, b_{j+1}\})$.

If $f(\{b_j, b_{j+1}\}) = (b_j, b_{j+1})$, let $u(b_j, b_{j+1}) = 1$.

If $f(\{b_j, b_{j+1}\}) = (b_{j+1}, b_j)$, let $u(b_{j+1}, b_j) = -1$.

With such a definition of u, it is immediate that Conditions (1)—(4) are satisfied. Thus, we need only verify Conditions (5)—(7).

To verify Condition (5) let $(s, t) \in$ dmn u and let $(t, m) \in$ dmn u. Suppose first that $(s, t) \neq (y, x)$ and $(t, m) \neq (y, x)$.

Then, for some $j \in [n]$, $(s, t) = (b_j, b_{j+1})$ or $(s, t) = (b_{j+1}, b_j)$.

If $(s, t) = (b_j, b_{j+1})$, then $j + 2 \in [n + 1]$ and
$(t, m) = (b_{j+1}, b_{j+2})$ and $u(s, t) = u(t, m) = 1$.

If $(s, t) = (b_{j+1}, b_j)$, then $j - 1 \in [n]$ and
$(t, m) = (b_j, b_{j-1})$ and $u(s, t) = u(t, m) = -1$.

Now, suppose that $(s, t) = (y, x)$ or $(t, m) = (y, x)$.

If $(s, t) = (y, x)$, then $(t, m) = (x, m) = (b_1, m) = (b_1, b_2)$,
and $u(s, t) = u(t, m) = 1$.

If $(t, m) = (y, x)$, then $(s, t) = (s, y) = (s, b_{n+1}) = (b_n, b_{n+1})$,
and $u(s, t) = u(t, m) = 1$.

To verify Condition (6) let $(s, t) \in$ dmn u and let $(m, t) \in$ dmn u with $(s, t) \neq (m, t)$.

Suppose first that $(s, t) \neq (y, x)$ and $(m, t) \neq (y, x)$.

Then, for some $j \in [n]$, $(s, t) = (b_j, b_{j+1})$ or $(s, t) = (b_{j+1}, b_j)$.

If $(s, t) = (b_j, b_{j+1})$, then $j + 2 \in [n + 1]$ and
$(m, t) = (b_{j+2}, b_{j+1})$ and $u(s, t) = 1$ and $u(m, t) = -1$.

If $(s, t) = (b_{j+1}, b_j)$ then $j - 1 \in [n]$ and
$(m, t) = (b_{j-1}, b_j)$ and $u(s, t) = -1$ and $u(m, t) = 1$.

Now, suppose that $(s, t) = (y, x)$ or $(m, t) = (y, x)$.

If $(s, t) = (y, x)$, then $(m, t) = (m, x) = (m, b_1) = (b_2, b_1)$, and
$u(s, t) = 1$ and $u(m, t) = -1$.

If $(m, t) = (y, x)$, then $(s, t) = (s, x) = (s, b_1) = (b_2, b_1)$, and
$u(s, t) = -1$ and $u(m, t) = 1$.

To verify Condition (7) let $(t, s) \in$ dmn u and let $(t, m) \in$ dmn u with $(t, s) \neq (t, m)$.

Suppose first that $(t, s) \neq (y, x)$ and $(t, m) \neq (y, x)$.

Then, for some $j \in [n]$, $(t, s) = (b_j, b_{j+1})$ or $(t, s) = (b_{j+1}, b_j)$.

If $(t, s) = (b_j, b_{j+1})$, then $j - 1 \in [n]$ and
$(t, m) = (b_j, b_{j-1})$ and $u(t, s) = 1$ and $u(t, m) = -1$.

If $(t, s) = (b_{j+1}, b_j)$ then $j + 2 \in [n + 1]$ and
$(t, m) = (b_{j+1}, b_{j+2})$ and $u(t, s) = -1$ and $u(t, m) = 1$.

Now, suppose that $(t, s) = (y, x)$ or $(t, m) = (y, x)$.

If $(t, s) = (y, x)$, then $(t, m) = (y, m) = (b_{n+1}, m) = (b_{n+1}, b_n)$,
and $u(t, s) = 1$ and $u(t, m) = -1$.

If $(t, m) = (y, x)$, then $(t, s) = (y, s) = (b_{n+1}, x) = (b_{n+1}, b_n)$,
and $u(t, m) = 1$ and $u(t, s) = -1$.

The proof is complete.

Now, let T be a tree in (V, S) such that $\sigma T = V$ and let $\{x, y\} \in S - T$. We are prepared to define $c(T, \{x, y\})$ as an element of $\mathscr{L}(K)$.

Because of the existence of a function with the properties described in the preceding Lemma, we can specify $c(T, \{x, y\})$ as such an element of $\mathscr{L}(K)$ that maps each directed branch $(s, t) \in K$ with $\{s, t\} \notin \lambda(T, \{x, y\})$ into 0, and whose restriction to $\mathrm{rng}\,(f|\lambda(T, \{x, y\}))$ is a function satisfying the conditions of the preceding Lemma. Formally,

$c(T, \{x, y\})$ is such a function that

(1) $c(T, \{x, y\}) \in \mathscr{L}(K)$.

(2) $(s, t) \in K$ and $\{s, t\} \notin \lambda(T, \{x, y\})$
implies
$(c(T, \{x, y\}))\,(s, t) = 0$.

(3) $\mathrm{rng}\,(c(T, \{x, y\})|\mathrm{rng}\,(f|\lambda(T, \{x, y\}))) \subset \{1, -1\}$.

(4) $(c(T, \{x, y\}))\,(f(\{x, y\})) = 1$.

(5) $(s, t) \in \mathrm{rng}\,(f|\lambda(T, \{x, y\}))$
and $(t, m) \in \mathrm{rng}\,(f|\lambda(T, \{x, y\}))$
implies
$(c(T, \{x, y\}))\,(s, t) = (c(T, \{x, y\}))\,(t, m)$.

(6) $(s, t) \in \mathrm{rng}\,(f|\lambda(T, \{x, y\}))$
and $(m, t) \in \mathrm{rng}\,(f|\lambda(T, \{x, y\}))$ and $(s, t) \neq (m, t)$
implies
$(c(T, \{x, y\}))\,(s, t) = -(c(T, \{x, y\})\,(m, t)$.

(7) $(t, s) \in \mathrm{rng}\,(f|\lambda(T, \{x, y\}))$
and $(t, m) \in \mathrm{rng}\,(f|\lambda(T, \{x, y\}))$ and $(t, s) \neq (t, m)$
implies
$(c(T, \{x, y\}))\,(t, s) = -(c(T, \{x, y\}))\,(t, m)$.

Next, let T and $\{x, y\}$ be as described above. We show in the following Theorem 1, as previously advertised, that the function $c(T, \{x, y\})$ is indeed a cycle.

Theorem 1. *Let T be a tree in (V, S). Suppose that $\sigma T = V$. Let $\{x, y\} \in S - T$.*
Then,

$c(T, \{x, y\}) \in \mathrm{kernel}\ \Delta$.

Proof: For brevity, let $u = c(T, \{x, y\})$.
We must show that $\Delta(u) = \theta_0$.

Let $b \in V$.
It suffices to show that $(\Delta(u))(b) = 0$.

Now,

$$(\Delta(u))(b) = \left(\Delta\left(\sum_{(s,t)\in K} u(s,t)\,\Phi^K_{(s,t)}\right)\right)(b)$$
$$= \left(\sum_{(s,t)\in K} u(s,t)\,\Delta(\Phi^K_{(s,t)})\right)(b)$$
$$= \sum_{(s,t)\in K} u(s,t)\,(\Delta(\Phi^K_{(s,t)}))(b).$$

For brevity, let $M = \lambda(T, \{x, y\})$.
From the definition of $c(T, \{x, y\})$, we know that if $(s, t) \in K$ and $(s, t) \notin \mathrm{rng}\,(f|M)$, then $u(s, t) = 0$.

Thus,

$$(\Delta(u))(b) = \sum_{(s,t)\in K} u(s,t)\,(\Delta(\Phi^K_{(s,t)}))(b)$$
$$= \sum_{(s,t)\in \mathrm{rng}\,(f|M)} u(s,t)\,(\Delta(\Phi^K_{(s,t)}))(b)$$
$$= \sum_{(s,t)\in \mathrm{rng}\,(f|M)} u(s,t)\,(\Phi^V_t - \Phi^V_s)(b)$$
$$= \sum_{(s,t)\in \mathrm{rng}\,(f|M)} u(s,t)\,(\Phi^V_t(b) - \Phi^V_s(b)).$$

Suppose first that $b \notin \sigma M$.
Then, whenever $(s, t) \in \mathrm{rng}\,(f|M)$, we know that $s \neq b$ and $t \neq b$, implying that $\Phi^V_s(b) = 0$ and $\Phi^V_t(b) = 0$.
Thus, if $b \notin \sigma M$, then $(\Delta(u))(b) = 0$, and we are through.

We now assume that $b \in \sigma M$, implying that b is a vertex of the loop M.
Thus, $\deg_M(b) = 2$.
Produce m and w with $m \neq w$ such that $\{b, m\} \in M$ and $\{b, w\} \in M$.

Now, if $(s, t) \in \mathrm{rng}\,(f|M)$ and $(s, t) \neq f(\{b, m\})$ and $(s, t) \neq f(\{b, w\})$, then $s \neq b$ and $t \neq b$, implying that $\Phi^V_s(b) = 0$ and $\Phi^V_t(b) = 0$.

Thus,

$$(\Delta(u))(b) = \sum_{(s,t)\in \mathrm{rng}\,(f|M)} u(s,t)\,(\Phi^V_t(b) - \Phi^V_s(b))$$
$$= \sum_{(s,t)\in \mathrm{rng}\,(f|\{\{b,m\},\{b,w\}\})} u(s,t)\,(\Phi^V_t(b) - \Phi^V_s(b)).$$

Four cases arise.

Case 1. $f(\{b, m\}) = (m, b)$ and $f(\{b, w\}) = (b, w)$.

Then, by Condition (5) of the definition of $c(T, \{x, y\})$, $u(m, b) = u(b, w)$, and

$$(\Delta(u))(b) = u(m,b)(\Phi_b^V(b) - \Phi_m^V(b)) + u(b,w)(\Phi_w^V(b) - \Phi_b^V(b))$$
$$= u(m,b) - u(b,w)$$
$$= 0.$$

Case 2. $f(\{b,m\}) = (m,b)$ and $f(\{b,w\}) = (w,b)$.

Then, by Condition (6) of the definition of $c(T,\{x,y\})$,
$u(m,b) = -u(w,b)$, and
$$(\Delta(u))(b) = u(m,b)(\Phi_b^V(b) - \Phi_m^V(b)) + u(w,b)(\Phi_b^V(b) - \Phi_w^V(b))$$
$$= u(m,b) + u(w,b)$$
$$= 0.$$

Case 3. $f(\{b,m\}) = (b,m)$ and $f(\{b,w\}) = (b,w)$.

Then, by Condition (7) of the definition of $c(T,\{x,y\})$,
$u(b,m) = -u(b,w)$, and
$$(\Delta(u))(b) = u(b,m)(\Phi_m^V(b) - \Phi_b^V(b)) + u(b,w)(\Phi_w^V(b) - \Phi_b^V(b))$$
$$= -u(b,m) - u(b,w)$$
$$= 0.$$

Case 4. $f(\{b,m\}) = (b,m)$ and $f(\{b,w\}) = (w,b)$.

Then, by Condition (5) of the definition of $c(T,\{x,y\})$,
$u(b,m) = u(w,b)$, and
$$(\Delta(u))(b) = u(b,m)(\Phi_m^V(b) - \Phi_b^V(b)) + u(w,b)(\Phi_b^V(b) - \Phi_w^V(b))$$
$$= -u(b,m) + u(w,b)$$
$$= 0.$$

Again, let T be a tree in (V,S) such that the vertices of T comprise all of the vertices of V and such that $T \neq S$. For each branch $\{x,y\} \in S - T$ we manufacture the cycle $c(T,\{x,y\})$. In the following Theorem 2 we show that the set of all such cycles $c(T,\{x,y\})$, each manufactured from the branch $\{x,y\} \in S - T$, is a base of the cycle space, kernel Δ. This base corresponds to the set of auxiliary loop current variables commonly used by the engineer in the process described previously, and it will be of use to us later in Chapter Eight.

Theorem 2. *Let (V,S) be a connected network. Let T be a tree in (V,S). Suppose that $\sigma T = V$. Suppose also that $T \neq S$.*
Then,

$\{c(T,\{x,y\}) \,|\, \{x,y\} \in S - T\}$ *is a base of* kernel Δ.

Proof: Let $B = \{c(T,\{x,y\}) \,|\, \{x,y\} \in S - T\}$.
We know from Theorem 1 that $B \subset$ kernel Δ.

But from the Corollary of 5.14, since (V,S) is a connected network

$$\dim(\text{kernel } \Delta) = \dim(\mathscr{L}(K)) - \dim(\mathscr{L}(V)) + 1$$
$$= pK - pV + 1$$
$$= pS - pV + 1.$$

Thus, to complete the proof it suffices to show that $pB = pS - pV + 1$ and B is linearly independent. This is done in the remainder of the proof which is divided into three parts.

Part 1. $\{x, y\} \in S - T$ and $\{s, t\} \in S - T$ and $\{x, y\} \neq \{s, t\}$ implies
$$(c(T, \{x, y\}))\, (f(\{s, t\})) = 0.$$

Proof of Part 1: Let $\{x, y\} \in S - T$ and let $\{s, t\} \in S - T$ with $\{x, y\} \neq \{s, t\}$. We know that $\lambda(T, \{x, y\}) \subset T \cup \{\{x, y\}\}$.
Thus, since $\{x, y\} \neq \{s, t\}$, we conclude that $\{s, t\} \notin \lambda(T, \{x, y\})$. By Condition (2) of the definition of $c(T, \{x, y\})$, $(c(T, \{x, y\}))\, (f(\{s, t\})) = 0.$

Part 2. $pB = pS - pV + 1.$

Proof of Part 2: By Theorem 1 of 2.10,
$$\begin{aligned}
p(S - T) &= pS - pT \\
&= pS - (p\sigma T - 1) \\
&= pS - p\sigma T + 1 \\
&= pS - pV + 1.
\end{aligned}$$

Thus, it suffices to show that $pB = p(S - T)$.

It is immediate that $pB \leq p(S - T)$.
To show that $pB = p(S - T)$ let $\{x, y\} \in S - T$ and let $\{s, t\} \in S - T$ with $\{x, y\} \neq \{s, t\}$. It suffices to show that $c(T, \{x, y\}) \neq c(T, \{s, t\})$.
But by Part 1, $(c(T, \{x, y\}))\, (f(\{s, t\})) = 0.$
By Condition (4) in the definition of $c(T, \{s, t\})$,
$(c(T, \{s, t\}))\, (f(\{s, t\})) = 1.$
Thus, $c(T, \{x, y\}) \neq c(T, \{s, t\}).$

Part 3. B is linearly independent.

Proof of Part 3: Let u be a function with dmn $u = S - T$ and rng $u \subset R$ such that
$$\sum_{\{x, y\} \in S - T} u(\{x, y\})\, c(T, \{x, y\}) = \theta_1.$$

It suffices to show that $u(\{s, t\}) = 0$ for each $\{s, t\} \in S - T$.

Let $\{s, t\} \in S - T$.
Then, by Part 1 and Condition (4) of the definition of $c(T, \{s, t\})$,
$$\begin{aligned}
0 &= \theta_1(f(\{s, t\})) \\
&= \Big(\sum_{\{x, y\} \in S - T} u(\{x, y\})\, c(T, \{x, y\})\Big)\, (f(\{s, t\})) \\
&= \sum_{\{x, y\} \in S - T} u(\{x, y\})\, (c(T, \{x, y\}))\, (f(\{s, t\})) \\
&= u(\{s, t\})\, (c(T, \{s, t\}))\, (f(\{s, t\})) \\
&= u(\{s, t\}).
\end{aligned}$$

Again, let T be a tree in (V, S) such that the vertices of T comprise all of the vertices of V and let $\{x, y\} \in S - T$. We can manufacture a loop $\lambda(T, \{x, y\})$ yielding an auxiliary loop current variable, say j, and we can also manufacture the corresponding cycle $c(T, \{x, y\})$. To conclude this Paragraph we discuss another relationship between the auxiliary loop current variable j and the cycle $c(T, \{x, y\})$. This relationship will be useful in the following Paragraph 7.6, when we discuss the dual situation involving auxiliary cut set voltage variables.

Now, suppose that $\{s, t\}$ is any branch of S and let $f(\{s, t\}) = (s, t)$. Suppose also that $\{s, t\}$ is contained in some loop of the network. We know that the value of the branch current $i(s, t)$ can be written as a sum of auxiliary loop current variables. More precisely, $i(s, t)$ is the sum of those auxiliary loop current variables corresponding to loops which contain the branch $\{s, t\}$. Thus, those directed branches which will have our particular auxiliary loop current variable j appearing in the expansion of the branch current as a sum of auxiliary loop current variables are precisely those directed branches which, undirected, are contained in the loop $\lambda(T, \{x, y\})$.

On the other hand, Conditions (2) and (3) of the definition of $c(T, \{x, y\})$ guarantee that $(c(T, \{x, y\}))\,(s, t) \neq 0$ if and only if $\{s, t\}$ is a branch of the loop $\lambda(T, \{x, y\})$.

Thus, we conclude that those directed branches which will have our particular auxiliary loop current variable j appearing in the expansion of the branch current as a sum of auxiliary loop current variables are precisely those directed branches to which the cycle $c(T, \{x, y\})$ assigns a non-zero value.

7.6 Voltage Variables

We now consider the situation dual to that of Paragraph 7.5. Assume that an electrical engineer is given a reasonably simple connected resistive network energized exclusively by current sources inducing a current chain i and a voltage chain v satisfying Ohm's Law and Kirchhoff's Voltage Law and Kirchhoff's Current Law. He is required to find the branch voltage $v(x, y)$ for each directed branch (x, y) of the network.

In such a situation the engineer usually does not try to compute directly the value of the voltage chain $v(x, y)$ for each directed branch (x, y) of the network. Instead, he introduces some auxiliary voltage variables called cut set voltage variables. Specifically, certain cut sets in the network are selected, and then a voltage is assigned to each cut set by a process which we shall subsequently describe.

Throughout the entire discussion involving these voltage variables the reader is urged to refer back to the discussion of Paragraph 7.5 involving current variables, so that the duality of the situation may be exposed. We should also warn the reader that there is confusion in the literature regarding the names assigned to these auxiliary voltage variables. Some writters incorrectly refer to these auxiliary voltage variables as node pair voltage variables. However, the node pair is an entirely different concept which we have no need to consider in this book.

The cut sets selected for the auxiliary cut set voltage variables must be selected with some care. The engineer must select a set of cut sets with the property that each cut set of the set contains a designated branch which is not contained in any other cut set of the set. In each such cut set, he considers the value of the branch voltage of this designated branch, and agrees that the cut set voltage of that cut set will be set equal to the value of this designated branch voltage. In this way, a set of auxiliary cut set voltage variables is manufactured.

A set of such cut sets with the property that each cut set of the set contains a designated branch which is not contained in any other cut set of the set is usually called an independent set of cut sets. The word "independent" is used to suggest that no member of the set of cut sets can be manufactured from the remaining cut sets of the set by some vague process involving the combining of branches from the remaining cut sets of the set.

The engineer must also select exactly the right number of auxiliary cut set voltage variables to be used in the analysis. Since each independent cut set gives rise to one auxiliary cut set voltage variable, he must specify the number of independent cut sets to be selected. This magic number is usually specified as the maximal number of such independent cut sets which can be produced from the network.

A maximal set of independent cut sets has another property called completeness. By completeness, we mean that each branch voltage can be written as a sum of the auxiliary cut set voltage variables. More precisely, each branch is contained in some of the independent cut sets, and the branch voltage of each branch is the sum of the auxiliary cut set voltage variables of those cut sets of the independent set of cut sets which contain the branch in question. Thus, to determine all branch voltages it suffices to find the value of each of the auxiliary cut set voltage variables.

Instead of seeking a maximal set of independent cut sets, the engineer could look for a set of independent cut sets which is also complete. Such a set of cut sets is automatically a maximal set of independent cut sets. Alternatively, he could look for a minimal set of complete cut sets. Such a set would automatically be an independent set of cut sets. We assume that by any of these processes the engineer has established an appropriate set of cut sets and corresponding auxiliary cut set voltage variables.

In each cut set of the maximal independent set of cut sets, the engineer now selects the vertices appearing in this cut set, and applies Kirchhoff's Current Law to this selected set of vertices, expressing each branch voltage involved in terms of the auxiliary cut set voltage variables. In this way the engineer obtains a system of linear equations in these auxiliary cut set voltage variables. There are precisely as many linear equations in this system as there are auxiliary cut set voltage variables. This system of linear equations is then solved to yield the values of the auxiliary cut set voltage variables. The branch voltages can then be obtained from the values of the auxiliary cut set voltage variables as described above.

Although the selection of an appropriate set of auxiliary cut set voltage variables appears to be a difficult job, it is a remarkable fact that the practicing engineer can usually apply this process to analyze a reasonably simple network and obtain the values of the branch voltages. For such simple networks the engineer instinctively knows how to select an appropriate maximal independent, or minimal complete set of cut sets for the auxiliary cut set voltage variables. It is our intention here to show that this popular process involving the selection of auxiliary cut set voltage variables is soundly based upon precise principles which we can derive from our abstract formulation.

In our abstract formulation, the precise analogue of a cut set is a coboundary. An independent set of cut sets translates into a linearly independent set of coboundaries, while a complete set of cut sets translates into a generating set of coboundaries. Thus, a set of cut sets which is both independent and complete has as its precise analogue a linearly independent generating set of coboundaries, which is, of course, a base for the coboundary space, range δ. Alternatively, a maximal independent set of cut sets corresponds to a maximal linearly independent set of coboundaries, which is a base, while a minimal complete set of cut sets corresponds to a minimal generating set of coboundaries, which is a base. Thus, the precise analogue of a set of cut sets appropriate for the auxiliary cut set voltage variables is merely a base of the coboundary space, range δ.

The number of auxiliary cut set voltage variables is equal to the number of elements of any base of the coboundary space, range δ, and this number is equal to the dimension of the coboundary space, range δ. We remind the reader that we have investigated the dimension of range δ in Chapter 5. In particular, we showed in Corollary 1 of 5.13 that

$$\dim (\mathrm{rng}\ \delta) = \dim (\mathrm{rng}\ \Delta).$$

Thus, applying the Theorem of 4.9 and the Theorem of 5.12,

$$\dim (\mathscr{L}(V)) = \dim (\mathrm{kernel}\ \delta) + \dim (\mathrm{rng}\ \delta).$$

Hence, under the assumption that (V, S) is a connected network, Theorem 3 of 5.14 gives

$$\dim (\mathrm{rng}\ \delta) = \dim (\mathscr{L}(V)) - 1 = pV - 1.$$

Thus, when (V, S) is a connected network, the number of auxiliary cut set voltage variables is one less than the number of vertices of V.

Since our unknown voltage chain v is a coboundary, we can express v is a linear combination of any base of the coboundary space, range δ. This fact is the precise analogue of our earlier assertion that each branch voltage can be written as a linear combination of the auxiliary cut set voltage variables.

Now, there are many different bases for the coboundary space, range δ. Thus, the engineer could select many different sets of auxiliary cut set voltage variables to be used in his analysis. We shall describe one process which the engineer uses to produce an appropriate set of auxiliary cut set voltage variables. Then we shall show that a precise analogue of this process can be used to construct a base of the coboundary space, range δ. In fact, the particular base of the coboundary space, range δ, that we construct, will be of use to us in our work in the following Chapter Eight.

For the remainder of this Paragraph we assume that (V, S) is a connected network. We exhibit first a process used by the engineer to manufacture an appropriate set of auxiliary cut set voltage variables.

Let T be a tree in (V, S) and suppose that each vertex of V is a vertex of T. Consider any branch $\{s, t\}$ of T. Apply the Theorem of 2.15 to this branch $\{s, t\}$ to produce a cut set M such that $\{s, t\}$ is a branch of M, but M includes no other branches of the tree T, and such that a branch $\{x, y\} \in S - T$ is an element of M if and only if $\{s, t\}$ is an element of $\lambda(T, \{x, y\})$, the unique loop including $\{x, y\}$ and contained in $T \cup \{\{x, y\}\}$.

For each branch $\{s, t\}$ of T, manufacture such a cut set M. Clearly, all such manufactured cut sets will be distinct. There will be as many such manufactured cut sets as there are branches of T. But from Theorem 1 of 2.10, the number of branches of T is one less than the number of vertices of T. Our assumption that each vertex of V is a vertex of T means that the vertices of T are precisely the vertices of V. Thus, the number of branches of T is one less than the number of vertices of V, and the number of such manufactured cut sets is

$$pV - 1,$$

which is the magic number of cut sets needed by the engineer to establish an appropriate set of auxiliary cut set voltage variables. Furthermore, this set of $pV - 1$ manufactured cut sets is independent, in the sense demanded by the engineer in order to establish a set of auxiliary cut set voltage variables. Thus, by assigning a voltage variable to each of these $pV - 1$ manufactured cut sets the engineer obtains an appropriate set of auxiliary cut set voltage variables.

It is worthwhile to note that the process described above depends upon the tree T used. Thus, for each such tree T, an appropriate set of auxiliary cut set voltage variables can be manufactured. Therefore, there are at least as many different choices of auxiliary cut set voltage variables as there are trees whose vertices comprise all the vertices of V. However, it turns out that there are many more sets of auxiliary cut set voltage variables than those which can be manufactured from trees by the process described above.

We now exhibit a precise analogue of the preceding process to manufacture a base of the coboundary space, range δ.

Let T be a tree in (V, S) whose vertices comprise all of the vertices of V. Thus, $\sigma T = V$. As before, for each branch $\{s, t\}$ of T, we apply the Theorem of 2.15 to the branch $\{s, t\}$ to produce a cut set including $\{s, t\}$ as a branch, but including no other branches of T, and such that a branch $\{x, y\} \in S - T$ is an element of this cut set if and only if $\{s, t\}$ is an element of $\lambda(T, \{x, y\})$, the unique loop including $\{x, y\}$ and contained in $T \cup \{\{x, y\}\}$. Throughout the remainder of this book we shall have a need for this cut set and thus, we adopt $\mu(T, \{s, t\})$ as a symbol for this cut set. Formally,

$\mu(T, \{s, t\})$ is such that

(1) $\mu(T, \{s, t\})$ is a cut set in (V, S).

(2) $\mu(T, \{s, t\}) \cap T = \{\{s, t\}\}$.

(3) $\{x, y\} \in S - T$
      implies
      $\{x, y\} \in \mu(T, \{s, t\})$ if and only if $\{s, t\} \in \lambda(T, \{x, y\})$.

Consider some branch $\{s, t\}$ of T and the cut set $\mu(T, \{s, t\})$. From this cut set we want to manufacture a coboundary which will be an element of our base of the coboundary space, range δ. This coboundary to be manufactured from $\mu(T, \{s, t\})$ must be an element of $\mathscr{L}(K)$, and thus, it must be a function whose domain is equal to K and whose range consists of real numbers. As a first step in the specification of such a function it seems reasonable to assign the number 0 to each directed branch (x, y) of K such that $\{x, y\}$ is not an element of $\mu(T, \{s, t\})$.

Thus, we must decide how this function will assign numbers to those directed branches (x, y) of K such that $\{x, y\}$ is an element of $\mu(T, \{s, t\})$. To each such directed branch (x, y) we shall assign either the number 1 or the number -1 as specified below. Our rule for the assignment of the number 1 or -1 may appear to be arbitrary and capricious, but we shall show subsequently in Theorem 1 that such an assignment of the number 1 or -1 does indeed result in the specification of a coboundary.

Let (x, y) be a directed branch of K such that $\{x, y\}$ is an element of $\mu(T, \{s, t\})$. Suppose first that $f(\{s, t\}) = (s, t)$, and consider $\operatorname{star}^V_{T-\{\{s,t\}\}}(s)$. Recall the Theorem of 2.16 which guarantees that either $x \in \operatorname{star}^V_{T-\{\{s,t\}\}}(s)$ and $y \notin \operatorname{star}^V_{T-\{\{s,t\}\}}(s)$, or $x \notin \operatorname{star}^V_{T-\{\{s,t\}\}}(s)$ and $y \in \operatorname{star}^V_{T-\{\{s,t\}\}}(s)$. If $x \in \operatorname{star}^V_{T-\{\{s,t\}\}}(s)$ and $y \notin \operatorname{star}^V_{T-\{\{s,t\}\}}(s)$, assign the number -1 to the directed branch (x, y). If $x \notin \operatorname{star}^V_{T-\{\{s,t\}\}}(s)$ and $y \in \operatorname{star}^V_{T-\{\{s,t\}\}}(s)$, assign the number 1 to the directed branch (x, y).

Now, suppose that $f(\{s, t\}) = (t, s)$, and consider $\operatorname{star}^V_{T-\{\{s,t\}\}}(t)$. By the Theorem of 2.16, either $x \in \operatorname{star}^V_{T-\{\{s,t\}\}}(t)$ and $y \notin \operatorname{star}^V_{T-\{\{s,t\}\}}(t)$, or $x \notin \operatorname{star}^V_{T-\{\{s,t\}\}}(t)$ and $y \in \operatorname{star}^V_{T-\{\{s,t\}\}}(t)$. If $x \in \operatorname{star}^V_{T-\{\{s,t\}\}}(t)$ and $y \notin \operatorname{star}^V_{T-\{\{s,t\}\}}(t)$, assign the number -1 to the directed branch (x, y). If $x \notin \operatorname{star}^V_{T-\{\{s,t\}\}}(t)$ and $y \in \operatorname{star}^V_{T-\{\{s,t\}\}}(t)$, assign the number 1 to the directed branch (x, y).

Thus, according to the recipe specified above, we have manufactured an element of $\mathscr{L}(K)$, and we show in Theorem 1 that this manufactured element of $\mathscr{L}(K)$ is indeed a coboundary. Before proceeding, however, it is convenient to adopt a special symbol for the element of $\mathscr{L}(K)$ specified above, and we adopt the symbol $d(T, \{s, t\})$ for this function. Formally,

$d(T, \{s, t\})$ is such a function that

(1) $d(T, \{s, t\}) \in \mathscr{L}(K)$.

(2) $(x, y) \in K$ and $\{x, y\} \notin \mu(T, \{s, t\})$
implies
$(d(T, \{s, t\}))\,(x, y) = 0$.

(3) $(x, y) \in K$ and $\{x, y\} \in \mu(T\ \{s, t\})$ and
$f(\{s, t\}) = (s, t)$ and
$x \in \operatorname{star}^V_{T-\{\{s,t\}\}}(s)$ and $y \notin \operatorname{star}^V_{T-\{\{s,t\}\}}(s)$
implies
$(d(T, \{s, t\}))\,(x, y) = -1$.

(4) $(x, y) \in K$ and $\{x, y\} \in \mu(T, \{s, t\})$ and
$f(\{s, t\}) = (s, t)$ and
$x \notin \operatorname{star}^V_{T-\{\{s,t\}\}}(s)$ and $y \in \operatorname{star}^V_{T-\{\{s,t\}\}}(s)$
implies
$(d(T, \{s, t\}))\,(x, y) = 1$.

(5) $(x, y) \in K$ and $\{x, y\} \in \mu(T, \{s, t\})$ and
$f(\{s, t\}) = (t, s)$ and
$x \in \text{star}^V_{T - \{\{s, t\}\}}(t)$ and $y \notin \text{star}^V_{T - \{\{s, t\}\}}(t)$
implies
$(d(T, \{s, t\}))(x, y) = -1$.

(6) $(x, y) \in K$ and $\{x, y\} \in \mu(T, \{s, t\})$ and
$f(\{s, t\}) = (t, s)$ and
$x \notin \text{star}^V_{T - \{\{s, t\}\}}(t)$ and $y \in \text{star}^V_{T - \{\{s, t\}\}}(t)$
implies
$(d(T, \{s, t\}))(x, y) = 1$.

Now, let T and $\{s, t\}$ be as described above. We show in the following Theorem 1, as previously advertised, that the function $d(T, \{s, t\})$ is indeed a coboundary.

Theorem 1. *Let (V, S) be a connected network. Let T be a tree in (V, S). Suppose that $\sigma T = V$. Let $\{s, t\} \in T$.*
Then,

$d(T, \{s, t\}) \in \text{rng } \delta$.

Proof: We lose no generality by assuming that $f(\{s, t\}) = (s, t)$. Let $B = \text{star}^V_{T - \{\{s, t\}\}}(s)$.
Observe that

$$\sum_{b \in B} \delta(\Phi_b^V) \in \text{rng } \delta.$$

To complete the proof, we shall show that

$$d(T, \{s, t\}) = \sum_{b \in B} \delta(\Phi_b^V).$$

Thus, let $(x, y) \in K$.
It suffices to show that

$$(d(T, \{s, t\}))(x, y) = \left(\sum_{b \in B} \delta(\Phi_b^V) \right)(x, y).$$

Examine $\left(\sum_{b \in B} \delta(\Phi_b^V) \right)(x, y)$.

$$\left(\sum_{b \in B} \delta(\Phi_b^V) \right)(x, y) = \left(\sum_{b \in B} \left(\sum_{u \in \text{rng}((\text{inv } K)|\{b\})} \Phi_{(u, b)}^K - \sum_{u \in \text{rng}(K|\{b\})} \Phi_{(b, u)}^K \right) \right)(x, y)$$

$$= \left(\sum_{b \in B} \sum_{u \in \text{rng}((\text{inv } K)|\{b\})} \Phi_{(u, b)}^K - \sum_{b \in B} \sum_{u \in \text{rng}(K|\{b\})} \Phi_{(b, u)}^K \right)(x, y)$$

$$= \sum_{b \in B} \sum_{u \in \text{rng}((\text{inv } K)|\{b\})} \Phi_{(u, b)}^K(x, y) - \sum_{b \in B} \sum_{u \in \text{rng}(K|\{b\})} \Phi_{(b, u)}^K(x, y).$$

Two cases arise:

Case 1. $\{x, y\} \notin \mu(T, \{s, t\})$.

In this case, Condition (2) of the definition of $d(T, \{s, t\})$ guarantees that $(d(T, \{s, t\}))(x, y) = 0$.

Also, reference to the Theorem of 2.16 shows that the assumption $\{x, y\} \notin \mu(T, \{s, t\})$ implies that either $x \notin B$ and $y \notin B$, or $x \in B$ and $y \in B$.

If $x \notin B$ and $y \notin B$, then for each $b \in B$, $\Phi^K_{(u, b)}(x, u) = 0$ when $u \in \mathrm{rng}\,((\mathrm{inv}\,K)|\{b\})$, and $\Phi^K_{(b, u)}(x, u) = 0$ when $u \in \mathrm{rng}\,(K|\{b\})$, and thus,

$$\big(\sum_{b \in B} \delta(\Phi^V_b)\big)(x, y) = 0,$$

and we are through.

If $x \in B$ and $y \in B$, then

$$\big(\sum_{b \in B} \delta(\Phi^V_b)\big)(x, y) = \sum_{b \in B}\ \sum_{u \in \mathrm{rng}\,((\mathrm{inv}\,K)|\{b\})} \Phi^K_{(u, b)}(x, y) - \sum_{b \in B}\ \sum_{u \in \mathrm{rng}\,(K|\{b\})} \Phi^K_{(b, u)}(x, y)$$
$$= \Phi^K_{(x, y)}(x, y) - \Phi^K_{(x, y)}(x, y)$$
$$= 0,$$

and we are through.

Case 2. $\{x, y\} \in \mu(T, \{s, t\})$.

In this case, by the Theorem of 2.16, either $x \in B$ and $y \notin B$, or $x \notin B$ and $y \in B$.

If $x \in B$ and $y \notin B$, then by Condition (3) of the definition of $d(T, \{s, t\})$, $\big(d(T, \{s, t\})\big)(x, y) = -1$.

Also,

$$\big(\sum_{b \in B} \delta(\Phi^V_b)\big)(x, y) = \sum_{b \in B}\ \sum_{u \in \mathrm{rng}\,((\mathrm{inv}\,K)|\{b\})} \Phi^K_{(u, b)}(x, y) - \sum_{b \in B}\ \sum_{u \in \mathrm{rng}\,(K|\{b\})} \Phi^K_{(b, u)}(x, y)$$
$$= - \Phi^K_{(x, y)}(x, y)$$
$$= -1.$$

If $x \notin B$ and $y \in B$, then by Condition (4) of the definition of $d(T, \{s, t\})$, $\big(d(T, \{s, t\})\big)(x, y) = 1$.

Also,

$$\big(\sum_{b \in B} \delta(\Phi^V_b)\big)(x, y) = \sum_{b \in B}\ \sum_{u \in \mathrm{rng}\,((\mathrm{inv}\,K)|\{b\})} \Phi^K_{(u, b)}(x, y) - \sum_{b \in B}\ \sum_{u \in \mathrm{rng}\,(K|\{b\})} \Phi^K_{(b, u)}(x, y)$$
$$= \Phi^K_{(x, y)}(x, y)$$
$$= 1.$$

The proof is complete.

Again, let T be a tree in (V, S) such that the vertices of T comprise all of the vertices of V. For each branch $\{s, t\}$ of T we manufacture the coboundary $d(T, \{s, t\})$. In the following Theorem 2 we show that the set of all such coboundaries $d(T, \{s, t\})$, each manufactured from the branch $\{s, t\}$ of T, is a base of the coboundary space, range δ. This base corresponds to the set of auxiliary cut set voltage variables commonly used by the engineer in the process described previously, and it will be of use to us later in Chapter Eight.

Theorem 2. *Let (V, S) be a connected network. Let T be a tree in (V, S). Suppose that $\sigma T = V$.*
Then,

$\{d(T, \{s, t\}) \,|\, \{s, t\} \in T\}$ *is a base of* rng δ.

Proof: Let $B = \{d(T, \{s, t\}) \,|\, \{s, t\} \in T\}$.
We know from Theorem 1 that $B \subset$ rng δ.

But from Corollary 1 of 5.13, $\dim (\text{rng } \delta) = \dim (\text{rng } \Delta)$. Applying the Theorem of 4.9 and the Theorem of 5.12,

$$\dim (\mathscr{L}(V)) = \dim (\text{kernel } \delta) + \dim (\text{rng } \delta).$$

Hence, under the assumption that (V, S) is a connected network, Theorem 3 of 5.14 gives

$$\dim (\text{rng } \delta) = \dim (\mathscr{L}(V)) - 1 = pV - 1.$$

To complete the proof it suffices to show that $pB = pV - 1$ and B is linearly independent. This is accomplished in the remainder of the proof which is divided into three parts.

Part 1. $\{x, y\} \in T$ and $\{s, t\} \in T$ and $\{x, y\} \neq \{s, t\}$
implies
$(d(T, \{s, t\})) \, (f(\{x, y\})) = 0.$

Proof of Part 1: Let $\{x, y\} \in T$ and let $\{s, t\} \in T$ with $\{x, y\} \neq \{s, t\}$.
Now, we know from Condition (2) of the definition of $\mu(T, \{s, t\})$ that $\mu(T, \{s, t\}) \cap T = \{\{s, t\}\}$.
Thus, since $\{x, y\} \in T$ and $\{x, y\} \neq \{s, t\}$, $\{x, y\} \notin \mu(T, \{s, t\})$.
By Condition (2) of the definition of $d(T, \{s, t\})$,
$(d(T, \{s, t\}))(f(\{x, y\})) = 0.$

Part 2. $pB = pV - 1.$

Proof of Part 2: By Theorem 1 of 2.10, $pV - 1 = p\sigma T - 1 = pT$.
Thus, it suffices to show that $pB = pT$.

It is immediate that $pB \leq pT$.
To show that $pB = pT$, let $\{s, t\} \in T$ and $\{x, y\} \in T$ with $\{s, t\} \neq \{x, y\}$.
It suffices to show that $d(T, \{s, t\}) \neq d(T, \{x, y\})$.

But by Part 1, $(d(T, \{s, t\})) \, (f(\{x, y\})) = 0.$
On the other hand, since $\{x, y\} \in \mu(T, \{x, y\})$, Conditions (3)–(6) of the definition of $d(T, \{x, y\})$ guarantee that $|(d(T, \{x, y\})) \, (f(\{x, y\}))| = 1$.
Thus, $d(T, \{s, t\}) \neq d(T, \{x, y\})$.

Part 3. B is linearly independent.

Proof of Part 3: Let u be a function with dmn $u = T$ and rng $u \subset R$ such that

$$\sum_{\{s, t\} \in T} u(\{s, t\}) \, d(T, \{s, t\}) = \theta_1.$$

It suffices to show that $u(\{x, y\}) = 0$ for each $\{x, y\} \in T$.

Let $\{x, y\} \in T$.
Then, by Part 1,

$$0 = \theta_1(f(\{x, y\}))$$
$$= (\sum_{\{s,t\} \in T} u(\{s, t\}) \, d(T, \{s, t\})) \, (f(\{x, y\}))$$
$$= \sum_{\{s,t\} \in T} u(\{s, t\}) \, (d(T, \{s, t\})) \, (f(\{x, y\}))$$
$$= u(\{x, y\}) \, (d(T, \{x, y\})) \, (f(\{x, y\})).$$

But Conditions (3)—(6) of the definition of $d(T, \{x, y\})$ guarantee that $|(d(T, \{x, y\})) \, (f(\{x, y\}))| = 1$.
Thus, $u(\{x, y\}) = 0$.

We conclude this Paragraph with comments dual to those remarks with which we concluded the preceding Paragraph 7.5.

Let T be a tree in (V, S) such that the vertices of T comprise all of the vertices of V and let $\{s, t\} \in T$. We can manufacture a cut set $\mu(T, \{s, t\})$ yielding an auxiliray cut set voltage variable, say e, and we can also manufacture the corresponding coboundary $d(T, \{s, t\})$.

Now, suppose that $\{x, y\}$ is any branch of S and let $f(\{x, y\}) = (x, y)$. We know that the value of the branch voltage $v(x, y)$ can be written as a sum of auxiliary cut set voltage variables. More precisely, $v(x, y)$ is the sum of those auxiliary cut set voltage variables corresponding to cut sets which contain the branch $\{x, y\}$. Thus, those directed branches which will have our particular auxiliary cut set voltage variable e appearing in the expansion of the branch voltage as a sum of auxiliary cut set voltage variables are precisely those directed branches which, undirected, are contained in the cut set $\mu(T, \{s, t\})$.

On the other hand, Conditions (2)—(6) of the definition of $d(T, \{s, t\})$ guarantee that $(d(T, \{s, t\}) \, (x, y) \neq 0$ if and only if $\{x, y\}$ is a branch of the cut set $\mu(T, \{s, t\})$.

Thus, we conclude that those directed branches which will have our particular auxiliary branch voltage variable e appearing in the expansion of the branch voltage as a sum of auxiliary cut set voltage variables are precisely those directed branches to which the coboundary $d(T, \{s, t\})$ assigns a non-zero value.

CHAPTER EIGHT

Kirchhoff's Third and Fourth Laws

8.0 Introduction

Consider a connected resistive network energized exclusively by current sources or exclusively by voltage sources. In Chapter Seven we have established the existence and uniqueness of a current chain and a voltage chain describing the performance of the network and satisfying Ohm's Law and Kirchhoff's Voltage Law and Kirchhoff's Current Law. However, we pointed out that the formulas developed in Chapter Seven for the current chain and voltage chain are not suitable for specific calculation of the branch currents and the branch voltages. In this Chapter we develop additional formulas for the current chain and voltage chain from which the branch currents and branch voltages may be readily calculated. We conclude this Chapter by showing that these formulas are essentially invariant under changes in the orientation of the branches of the network.

8.1 Assumptions of This Chapter

As in Chapter 6 and Chapter 7 we agree to let R be the field of scalars for our linear algebra.

As in Chapter 5 and Chapter 6 and Chapter 7 we fix an incidence function f, and agree to let

$$S = \operatorname{dmn} f,$$
$$V = \sigma\operatorname{dmn} f = \sigma S,$$
$$K = \operatorname{rng} f.$$

We also assume throughout this Chapter that (V, S) is a connected network.

As in 6.2 and Chapter 7 we fix a resistance function r.

Thus, we agree that

$$r \text{ is a function,}$$
$$\operatorname{dmn} r = S,$$
$$\operatorname{rng} r \subset \{x \mid x \in R \text{ and } x > 0\}.$$

The impedance function Z described in 6.4 will be of use to us. Thus, as in Chapter 7, we agree to fix a function Z such that

> Z is the unique linear map of $\mathscr{L}(K)$ into $\mathscr{L}(K)$
> such that for each $(x, y) \in K$,
> $Z(\Phi^K_{(x,y)}) = r(\{x, y\})\, \Phi^K_{(x,y)}.$

As in Lemma 1 of 7.4, we agree to fix a function Y such that

> Y is the unique linear map of $\mathscr{L}(K)$ into $\mathscr{L}(K)$
> such that for each $(x, y) \in K$,
> $Y(\Phi^K_{(x,y)}) = (r(\{x, y\}))^{-1}\, \Phi^K_{(x,y)}.$

We remind the reader that Lemma 1 of 7.4 guarantees that $Y = \operatorname{inv} Z$ and Z is univalent.

8.2 The Cycle Map

Let T be a tree in (V, S) such that the vertices of T comprise all of the vertices of V. We have need for a linear map of $\mathscr{L}(K)$ into $\mathscr{L}(K)$ which maps $\Phi^K_{(x,y)}$ into θ_1 when $(x, y) \in K$ such that $\{x, y\} \in T$, and which maps $\Phi^K_{(x,y)}$ into the cycle $c(T, \{x, y\})$ when $(x, y) \in K$ such that $\{x, y\} \in S - T$. For each such tree T we shall define such a linear map to be denoted by A_T.

Since we are now considering all trees T in (V, S) whose vertices comprise all of the vertices of V, we adopt the symbol $\mathscr{T}$ for the set of all such trees. Formally,

$$\mathscr{T} = \{T \mid T \text{ is a tree in } (V, S) \text{ and } \sigma T = V\}.$$

We observe that as a consequence of the Theorem of 2.13, we know that $\mathscr{T}$ is non-empty, because of our assumptions that (V, S) is a connected network and f is an incidence function with domain S.

Now, for each $T \in \mathscr{T}$, we can define the cycle map. Formally,

> For each $T \in \mathscr{T}$,
> A_T is the unique linear map of $\mathscr{L}(K)$ into $\mathscr{L}(K)$
> such that for each $(x, y) \in K$,
> (1) $A_T(\Phi^K_{(x,y)}) = \theta_1$, if $\{x, y\} \in T$.
> (2) $A_T(\Phi^K_{(x,y)}) = c(T, \{x, y\})$, if $\{x, y\} \in S - T$.

We now investigate properties of the cycle map A_T when $T \in \mathscr{T}$.

Suppose that $T \in \mathscr{T}$, and $(x, y) \in K$ such that $\{x, y\} \in S - T$. As our first property of the cycle map A_T, we show in the following Theorem 1 that $c(T, \{x, y\})$ is invariant under the map A_T.

Theorem 1. *Let $T \in \mathcal{T}$. Let $(x, y) \in K$. Suppose that $\{x, y\} \in S - T$. Then,*

$$A_T(c(T, \{x, y\})) = c(T, \{x, y\}).$$

Proof: Note that $c(T, \{x, y\}) = \sum_{(s,t) \in K} (c(T, \{x, y\}))\,(s, t)\,\Phi^K_{(s,t)}$.

Thus,

$$A_T(c(T, \{x, y\})) = A_T\Big(\sum_{(s,t) \in K} (c(T, \{x, y\}))\,(s, t)\,\Phi^K_{(s,t)} \Big)$$
$$= \sum_{(s,t) \in K} (c(T, \{x, y\}))\,(s, t)\,A_T(\Phi^K_{(s,t)}).$$

Now, if $(s, t) \in K$ such that $\{s, t\} \in T$, then $A_T(\Phi^K_{(s,t)}) = \theta_1$.

Thus,

$$A_T(c(T, \{x, y\})) = \sum_{(s,t) \in K} (c(T, \{x, y\}))\,(s, t)\,A_T(\Phi^K_{(s,t)})$$
$$= \sum_{(s,t)\,\in\,\mathrm{rng}\,(f|(S-T))} (c(T, \{x, y\}))\,(s, t)\,A_T(\Phi^K_{(s,t)}).$$

Now, suppose that $(s, t) \in \mathrm{rng}\,(f|(S - T))$ with $(s, t) \neq (x, y)$ and consider $(c(T, \{x, y\}))\,(s, t)$. Since $\lambda(T, \{x, y\}) \subset T \cup \{\{x, y\}\}$ and $\{s, t\} \in S - T$, we have $\{s, t\} \notin \lambda(T, \{x, y\})$, and by Condition (2) of the definition of $c(T, \{x, y\})$ in 7.5, $(c(T, \{x, y\}))\,(s, t) = 0$.

Thus,

$$A_T(c(T, \{x, y\})) = \sum_{(s,t)\,\in\,\mathrm{rng}\,(f|(S-T))} (c(T, \{x, y\}))\,(s, t)\,A_T(\Phi^K_{(s,t)})$$
$$= (c(T, \{x, y\}))\,(x, y)\,A_T(\Phi^K_{(x,y)}).$$

But by Condition (4) of the definition of $c(T, \{x, y\})$ in 7.5, $(c(T, \{x, y\}))\,(x, y) = 1$.

Thus,

$$A_T(c(T, \{x, y\})) = (c(T, \{x, y\}))\,(x, y)\,A_T(\Phi^K_{(x,y)})$$
$$= A_T(\Phi^K_{(x,y)})$$
$$= c(T, \{x, y\}).$$

Again, let $T \in \mathcal{T}$. Using Theorem 1 it is easy to show that each cycle of kernel Δ is invariant under the cycle map A_T. This fact is exhibited as the following Theorem 2.

Theorem 2. *Let $T \in \mathcal{T}$. Let $z \in$ kernel Δ. Then,*

$$A_T(z) = z.$$

Proof: Using the Corollary of 5.14 and Theorem 1 of 2.10,

$$\dim(\text{kernel } \Delta) = \dim(\mathcal{L}(K)) - \dim(\mathcal{L}(V)) + 1$$
$$= pK - pV + 1$$
$$= pS - p\sigma T + 1$$
$$= pS - pT.$$

Thus, if $S = T$, then $\dim(\text{kernel } \varDelta) = 0$ and $z = \theta_1$, implying that $A_T(z) = A_T(\theta_1) = \theta_1$.

We now assume that $S \neq T$.

By Theorem 2 of 7.5, $\{c(T, \{x, y\}) \mid \{x, y\} \in S - T\}$ is a base of kernel $\varDelta$.

Produce a function u such that $\operatorname{dmn} u = S - T$ and $\operatorname{rng} u \subset R$ and

$$z = \sum_{\{x, y\} \in S - T} u(\{x, y\}) \, c(T, \{x, y\}).$$

Thus, by Theorem 1,

$$
\begin{aligned}
A_T(z) &= A_T\Big(\sum_{\{x, y\} \in S - T} u(\{x, y\}) \, c(T, \{x, y\}) \Big) \\
&= \sum_{\{x, y\} \in S - T} u(\{x, y\}) \, A_T\big(c(T, \{x, y\})\big) \\
&= \sum_{\{x, y\} \in S - T} u(\{x, y\}) \, c(T, \{x, y\}) \\
&= z.
\end{aligned}
$$

Let $T \in \mathcal{T}$. From the definition of the cycle map A_T it is apparent that the range of the cycle map A_T is a subset of the cycle space, kernel $\varDelta$. From Theorem 2 it is easy to show that the range of the cycle map A_T is, in fact, equal to the cycle space, kernel $\varDelta$. We exhibit this fact as the following Theorem 3.

Theorem 3. *Let $T \in \mathcal{T}$.*
Then,

$\operatorname{rng} A_T = \text{kernel } \varDelta.$

Proof: Let $w \in \operatorname{rng} A_T$. Produce $u \in \mathcal{L}(K)$ such that $w = A_T(u)$. Then,

$$
\begin{aligned}
w &= A_T(u) \\
&= A_T\Big(\sum_{(x, y) \in K} u(x, y) \, \varPhi^K_{(x, y)} \Big) \\
&= \sum_{(x, y) \in K} u(x, y) \, A_T(\varPhi^K_{(x, y)}).
\end{aligned}
$$

Let $(x, y) \in K$.

If $\{x, y\} \in T$, then $A_T(\varPhi^K_{(x, y)}) = \theta_1 \in \text{kernel } \varDelta$, and $u(x, y) \, A_T(\varPhi^K_{(x, y)}) \in \text{kernel } \varDelta$.

If $\{x, y\} \notin S - T$, then $A_T(\varPhi^K_{(x, y)}) = c(T, \{x, y\}) \in \text{kernel } \varDelta$, and $u(x, y) \, A_T(\varPhi^K_{(x, y)}) \in \text{kernel } \varDelta$.

Thus, $w \in \text{kernel } \varDelta$, and $\operatorname{rng} A_T \subset \text{kernel } \varDelta$.

On the other hand, let $z \in \text{kernel } \varDelta$.

Then, by Theorem 2, $z = A_T(z) \in \operatorname{rng} A_T$.

Thus, kernel $\varDelta \subset \operatorname{rng} A_T$.

Again, let $T \in \mathcal{T}$. The last property of the cycle map that we exhibit here concerns $(A_T)^t$, the transpose of the cycle map A_T, which we agree to denote more simply by A_T^t.

In the following Theorem 4 we show that A_T^t, the transpose of the cycle map A_T, sends each coboundary into θ_1, the zero element of $\mathcal{L}(K)$.

Theorem 4. *Let* $T \in \mathcal{T}$.
Then,
$$\text{rng } (A_T^t | \text{rng } \delta) = \{\theta_1\}.$$

Proof: It suffices to show that $\text{rng } (A_T^t | \text{rng } \delta) \subset \{\theta_1\}$.

Let $z \in \text{rng } \delta$.

We shall show that $A_T^t(z) = \theta_1$.

But by Theorem 1 of 4.4, Theorem 2 of 4.5, and Theorem 2 of 4.7,

$$\begin{aligned}
A_T^t(z) &= \sum_{(x,y) \in K} (A_T^t(z))(x,y)\, \Phi_{(x,y)}^K \\
&= \sum_{(x,y) \in K} \langle A_T^t(z),\, \Phi_{(x,y)}^K \rangle\, \Phi_{(x,y)}^K \\
&= \sum_{(x,y) \in K} \langle z,\, A_T(\Phi_{(x,y)}^K) \rangle\, \Phi_{(x,y)}^K.
\end{aligned}$$

Let $(x, y) \in K$.

By Theorem 3, $A_T(\Phi_{(x,y)}^K) \in \text{kernel } \Delta$.

By Theorem 1 of 5.9, $\langle z,\, A_T(\Phi_{(x,y)}^K) \rangle = 0$.

Thus, $A_T^t(z) = \theta_1$.

8.3 The Chord Map

If T is a tree in (V, S), by a chord of T we mean a branch of $S - T$. From all trees $T \in \mathcal{T}$, and the chords of such trees, we shall manufacture a linear map of $\mathcal{L}(K)$ into $\mathcal{L}(K)$ and investigate its properties.
This linear map to be manufactured will be called the chord map and will be denoted by F.

Before defining the chord map F, however, we must define a function W whose domain is equal to $\mathcal{T}$. For each $T \in \mathcal{T}$ we shall let $W(T)$ be the product of the values $r(\{x, y\})$ of the resistance function r, evaluated only for each chord $\{x, y\}$ of $S - T$. For each $T \in \mathcal{T}$ we call $W(T)$ the tree chord product. Formally,

W is the function with domain $\mathcal{T}$ such that for each $T \in \mathcal{T}$,

$$W(T) = \prod_{\{x,y\} \in S - T} r(\{x, y\}).$$

We remind the reader that if $T \in \mathcal{T}$ such that $T = S$ and $S - T$ is empty, then our convention of 5.5 guarantees that $W(T) = 1$.

The chord map F is now defined as the sum, for each $T \in \mathscr{T}$, of all linear maps $W(T) A_T$. Formally,

$$F = \sum_{T \in \mathscr{T}} W(T) A_T.$$

The first property of the chord map F, which we exhibit in the following Theorem 1, is the fact that the transpose F^t, of the chord map F, sends each coboundary into θ_1, the zero element of $\mathscr{L}(K)$.

Theorem 1. rng $(F^t | rng \; \delta) = \{\theta_1\}$.

Proof: Clearly, F is a linear map of $\mathscr{L}(K)$ into $\mathscr{L}(K)$, and thus, F^t is a linear map of $\mathscr{L}(K)$ into $\mathscr{L}(K)$.
Hence, it suffices to show that rng $(F^t | rng \; \delta) \subset \{\theta_1\}$.

Let $z \in rng \, \delta$.
We shall show that $F^t(z) = \theta_1$.

But by Theorem 5 of 4.7 and Theorem 6 of 4.7,

$$F^t(z) = \left(\sum_{T \in \mathscr{T}} W(T) A_T \right)^t (z)$$

$$= \left(\sum_{T \in \mathscr{T}} W(T) A_T^t \right) (z)$$

$$= \sum_{T \in \mathscr{T}} W(T) A_T^t (z).$$

Let $T \in \mathscr{T}$.
By Theorem 4 of 8.2, $A_T^t(z) = \theta_1$, implying that $W(T) A_T^t(z) = \theta_1$.
Thus, $F^t(z) = \theta_1$.

The next fact about the chord map F, which we shall exhibit in Theorem 2, is the rather innocuous assertion that the superposition function $Z \circ F$, is equal to its transpose, $(Z \circ F)^t$. Strangely enough, this simple assertion is rather difficult to prove, and its validity is the fundametal key to the explicit formulas for the current chain and voltage chain which we shall develop.

The proof of the assertion that $Z \circ F$ is equal to its transpose, $(Z \circ F)^t$, requires a succession of preliminary Lemmas. Since each such Lemma refers to a rather special situation, we shall not bother to describe each Lemma in prose. Instead, we shall merely exhibit each Lemma without comment, together with its proof. Then, at the conclusion of the preliminary Lemmas we shall exhibit Theorem 2, stating that $Z \circ F$ is equal to its transpose, $(Z \circ F)^t$, and show in the proof of Theorem 2 how the various Lemmas are needed.

Lemma 1. *Let $T \in \mathcal{T}$. Let $\{x, y\} \in S - T$. Let $\{s, t\} \in \lambda(T, \{x, y\})$.*
Suppose that $\{s, t\} \neq \{x, y\}$. Let $Q = (T \cup \{\{x, y\}\}) - \{\{s, t\}\}$.
Then,

    (1) $\sigma Q = V$.

    (2) $Q \in \mathcal{T}$.

    (8) $\lambda(T, \{x, y\}) = \lambda(Q, \{s, t\})$.

Proof of (1): Clearly, it suffices to show that $V \subset \sigma Q$.

Let $b \in V$.

Since $T \in \mathcal{T}$, $V = \sigma T$, and there exists $m \in \sigma T$ such that $b \in \{b, m\} \in T$.
Now, if $\{b, m\} \neq \{s, t\}$, then

$$b \in \{b, m\} \in (T \cup \{\{x, y\}\}) - \{\{s, t\}\} = Q,$$

and we are through.

Thus, assume that $\{b, m\} = \{s, t\}$.

Then, $\{b, m\} \in \lambda(T, \{x, y\})$ and $\deg_{\lambda(T, \{x, y\})}(b) = 2$.

Produce w with $w \neq m$ such that $\{b, w\} \in \lambda(T, \{x, y\}) \subset T \cup \{\{x, y\}\}$.
But since $w \neq m$, $\{b, w\} \neq \{s, t\}$, and

$$b \in \{b, w\} \in (T \cup \{\{x, y\}\}) - \{\{s, t\}\} = Q.$$

Hence, $b \in \sigma Q$.

Proof of (2): Note that $\{s, t\} \in \lambda(T, \{x, y\}) \subset T \cup \{\{x, y\}\}$.
Thus, since $\{x, y\} \notin T$,

$$\begin{aligned}
pQ &= p((T \cup \{\{x, y\}\}) - \{\{s, t\}\}) \\
&= p(T \cup \{\{x, y\}\}) - 1 \\
&= (pT + 1) - 1 \\
&= pT.
\end{aligned}$$

But since $T \in \mathcal{T}$, $p\sigma T = pT + 1$ and $\sigma T = V$.
Hence, by (1),

$$p\sigma Q = pV = p\sigma T = 1 + pT = 1 + pQ.$$

Thus, by Theorem 2 of 2.10, it suffices to show that Q is a connected set of branches.

By the Theorem of 1.20, it suffices to show that Q is path connected.
But since $\sigma Q = V$, $(\sigma Q, Q)$ is a network.
Thus, to show that Q is path connected, we let $u \in \sigma Q$ and let $v \in \sigma Q$ with $u \neq v$.

We shall exhibit a path from u to v in Q.

Now, $u \in \sigma Q = V = \sigma T$ and $v \in \sigma Q = V = \sigma T$.
Thus, produce h such that h is a proper path from u to v in T.

If $\{s, t\} \notin \text{rng } h$, then

$$\text{rng } h \subset T - \{\{s, t\}\} \subset (T \cup \{\{x, y\}\}) - \{\{s, t\}\} = Q,$$

and we are thruogh.

Thus, we assume that $\{s, t\} \in \mathrm{rng}\ h$.

Let $n \in \omega$ such that $\mathrm{dmn}\ h = [n]$.

By the Theorem of 1.15, produce a univalent sequence z with $\mathrm{dmn}\ z = [n+1]$ such that

$$z_1 = u,$$
$$z_{n+1} = v,$$
$$h_j = \{z_j, z_{j+1}\} \text{ for each } j \in [n].$$

But we have assumed that $\{s, t\} \in \mathrm{rng}\ h$.

We lose no generality by assuming that for some $k \in [n]$, $s = z_k$ and $t = z_{k+1}$ and $h_k = \{z_k, z_{k+1}\}$.

Since $\{s, t\} \in \lambda(T, \{x, y\})$, apply Theorem 1 of 2.5 to produce q such that q is a proper path from s to t in $\lambda(T, \{x, y\}) - \{\{s, t\}\} \subset (T \cup \{\{x, y\}\}) - \{\{s, t\}\} = Q$.

Now, if $s = u$, then q is a proper path from u to t in Q. On the other hand, if $s \neq u$, then $k > 1$ and $h|[k-1]$ is a proper path from u to s in $T - \{\{s, t\}\} \subset (T \cup \{\{x, y\}\}) - \{\{s, t\}\} = Q$. By combining $h|[k-1]$ with q, which is a proper path from s to t in Q, we can easily produce g such that g is a proper path from u to t in Q.

Thus, in any case, we can produce g such that g is a proper path from u to t in Q.

Now, if $t = v$, then g is a proper path from u to v in Q and we are through.

On the other hand, if $t \neq v$, note that

$$\{(j, h_{k+j}) \mid j \in [n-k]\}$$

is a proper path from t to v in $T - \{\{s, t\}\} \subset (T \cup \{\{x, y\}\}) - \{\{s, t\}\} = Q$. Thus, by combining $\{(j, h_{k+j}) \mid j \in [n-k]\}$ with g, which is a proper path from u to t in Q, we can easily produce a path from u to v in Q.

Proof of (3): Note that

$$\{s, t\} \in \lambda(T, \{x, y\}) \subset T \cup \{\{x, y\}\}.$$

But

$$T \cup \{\{x, y\}\} = ((T \cup \{\{x, y\}\}) - \{\{s, t\}\}) \cup \{\{s, t\}\} = Q \cup \{\{s, t\}\}.$$

Thus,

$$\{s, t\} \in \lambda(T, \{x, y\}) \subset Q \cup \{\{s, t\}\}.$$

On the other hand,

$$\{s, t\} \in \lambda(Q, \{s, t\}) \subset Q \cup \{\{s, t\}\}.$$

Also, $\{s, t\} \notin Q$ and Q is a tree.

Since $\lambda(T, \{x, y\})$ is a loop and $\lambda(Q, \{s, t\})$ is a loop, the uniqueness stipulated in the Theorem of 2.12 guarantees that
$\lambda(T, \{x, y\}) = \lambda(Q, \{s, t\})$.

Lemma 2. *Let $T \in \mathcal{T}$. Let $\{x, y\} \in S - T$. Let g be a proper path from x to t in $\lambda(T, \{x, y\}) - \{\{x, y\}\}$. Let $\{s, t\} = g_{pg}$.*
Then,

     (1) $(x, y) \in K$ *and* $(s, t) \in K$ *implies* $\big(c(T, \{x, y\})\big)(s, t) = -1$.

     (2) $(x, y) \in K$ *and* $(t, s) \in K$ *implies* $\big(c(T, \{x, y\})\big)(t, s) = 1$.

     (3) $(y, x) \in K$ *and* $(s, t) \in K$ *implies* $\big(c(T, \{x, y\})\big)(s, t) = 1$.

     (4) $(y, x) \in K$ *and* $(t, s) \in K$ *implies* $\big(c(T, \{x, y\})\big)(t, s) = -1$.

Proof: For brevity, let $u = c(T, \{x, y\})$. The proof is by induction on pg.

Suppose first that w is a proper path from x to b in
$\lambda(T, \{x, y\}) - \{\{x, y\}\}$ with $pw = 1$.
Then, $\{x, b\} = w_1 = w_{pw}$.
We proceed to verify the four required conditions of this Lemma.

If $(x, y) \in K$ and $(x, b) \in K$, then Condition (7) of the definition of u in 7.5 guarantees that $u(x, b) = -u(x, y) = -1$.
If $(x, y) \in K$ and $(b, x) \in K$, then Condition (5) of the definition of u in 7.5 guarantees that $u(b, x) = u(x, y) = 1$.
If $(y, x) \in K$ and $(x, b) \in K$, then Condition (5) of the definition of u in 7.5 guarantees that $1 = u(y, x) = u(x, b)$.
If $(y, x) \in K$ and $(b, x) \in K$, then Condition (6) of the definition of u in 7.5 guarantees that $u(b, x) = -u(y, x) = -1$.

Next, let $k \in \omega$ such that whenever h is a proper path from x to e in $\lambda(T, \{x, y\}) - \{\{x, y\}\}$ with $ph = k$ and $\{z, e\} = h_{ph} = h_k$, the following conditions are true.

     (1) $(x, y) \in K$ and $(z, e) \in K$ implies $u(z, e) = -1$.

     (2) $(x, y) \in K$ and $(e, z) \in K$ implies $u(e, z) = 1$.

     (3) $(y, x) \in K$ and $(z, e) \in K$ implies $u(z, e) = 1$.

     (4) $(y, x) \in K$ and $(e, z) \in K$ implies $u(e, z) = -1$.

Suppose that, as in the statement of the Lemma, g is a proper path from x to t in $\lambda(T, \{x, y\}) - \{\{x, y\}\}$ with $pg = k + 1$ and
$\{s, t\} = g_{pg} = g_{k+1}$.
Let $h = g|[k]$.
Observe that h is a proper path from x to s in $\lambda(T, \{x, y\}) - \{\{x, y\}\}$ with $ph = k$, and $s \in h_k$. Let $h_k = \{m, s\}$, and note that $m \neq t$.

We proceed to verify the four conditions of this Lemma by considering four separate cases.

Case 1. $(x, y) \in K$ and $(s, t) \in K$.

In this case, if $(m, s) \in K$, Condition (1) of the inductive hypothesis guarantees that $u(m, s) = -1$, and by Condition (5) of the definition of u in 7.5, $-1 = u(m, s) = u(s, t)$.
On the other hand, if $(s, m) \in K$, Condition (2) of the inductive hypothesis guarantees that $u(s, m) = 1$, and by Condition (7) of the definition of u in 7.5, $u(s, t) = -u(s, m) = -1$.

Case 2. $(x, y) \in K$ and $(t, s) \in K$.

In this case, if $(m, s) \in K$, Condition (1) of the inductive hypothesis guarantees that $u(m, s) = -1$, and by Condition (6) of the definition of u in 7.5, $u(t, s) = -u(m, s) = 1$.
On the other hand, if $(s, m) \in K$, Condition (2) of the inductive hypothesis guarantees that $u(s, m) = 1$, and by Condition (5) of the definition of u in 7.5, $u(t, s) = u(s, m) = 1$.

Case 3. $(y, x) \in K$ and $(s, t) \in K$.

In this case, if $(m, s) \in K$, Condition (3) of the inductive hypothesis guarantees that $u(m, s) = 1$, and by Condition (5) of the definition of u in 7.5, $1 = u(m, s) = u(s, t)$.
On the other hand, if $(s, m) \in K$, Condition (4) of the inductive hypothesis guarantees that $u(s, m) = -1$, and by Condition (7) of the definition of u in 7.5, $u(s, t) = -u(s, m) = 1$.

Case 4. $(y, x) \in K$ and $(t, s) \in K$.

In this case, if $(m, s) \in K$, Condition (3) of the inductive hypothesis guarantees that $u(m, s) = 1$, and by Condition (6) of the definition of u in 7.5, $u(t, s) = u(m, s) = -1$.
On the other hand, if $(s, m) \in K$, Condition (4) of the inductive hypothesis guarantees that $u(s, m) = -1$, and by Condition (5) of the definition of u in 7.5, $u(t, s) = u(s, m) = -1$.

Lemma 3. *Let* $T \in \mathscr{T}$. *Let* $\{x, y\} \in S - T$. *Let* $\{s, t\} \in \lambda(T, \{x, y\})$. *Suppose that* $\{x, y\} \neq \{s, t\}$. *Let* $Q = (T \cup \{\{x, y\}\}) - \{\{s, t\}\}$. *Then,*

$$(c(T, \{x, y\})) \, (f(\{s, t\})) = (c(Q, \{s, t\})) \, (f(\{x, y\})).$$

Proof: By Theorem 1 of 2.5, produce g such that g is a proper path from x to y in $\lambda(T, \{x, y\}) - \{\{x, y\}\}$.
By Theorem 2 of 2.5, $\operatorname{rng} g = \lambda(T, \{x, y\}) - \{\{x, y\}\}$, and thus, $\{s, t\} \in \operatorname{rng} g$.

Let $m \in \omega$ such that $\operatorname{dmn} g = [m]$.
By the Theorem of 1.15, produce a univalent sequence b with $\operatorname{dmn} b = [m + 1]$ such that

$$b_1 = x,$$
$$b_{m+1} = y,$$
$$g_j = \{b_j, b_{j+1}\} \text{ for each } j \in [m].$$

Since $\{s, t\} \in \text{rng } g$, we lose no generality by assuming that $k \in [m]$ such that $s = b_k$ and $t = b_{k+1}$ and $g_k = \{s, t\} = \{b_k, b_{k+1}\}$.

Note that $g|[k]$ is a proper path from x to t in $\lambda(T, \{x, y\}) - \{\{x, y\}\}$, and since $p(g|[k]) = k$, $\{s, t\} = (g|[k])_k = g_k$.
Thus, Lemma 2 applies to give the following four conditions.

(1) $(x, y) \in K$ and $(s, t) \in K$
     implies
     $(c(T, \{x, y\}))\,(s, t) = -1.$

(2) $(x, y) \in K$ and $(t, s) \in K$
     implies
     $(c(T, \{x, y\}))\,(t, s) = 1.$

(3) $(y, x) \in K$ and $(s, t) \in K$
     implies
     $(c(T, \{x, y\}))\,(s, t) = 1.$

(4) $(y, x) \in K$ and $(t, s) \in K$
     implies
     $(c(T, \{x, y\}))\,(t, s) = -1.$

Now, by Lemma 1, (2), $Q \in \mathcal{T}$, and by Lemma 1, (3),
$\lambda(T, \{x, y\}) = \lambda(Q, \{s, t\}).$

Next, we exhibit a proper path h from t to x in $\lambda(Q, \{s, t\}) - \{\{s, t\}\}$ such that $h_{ph} = \{y, x\}.$

If $t = y$, merely let $h = \{(1, \{y, x\})\}$, and check easily that h is a proper path from t to x in $\lambda(Q, \{s, t\}) - \{\{s, t\}\}$ and $h_{ph} = \{y, x\}.$

On the other hand, if $t \neq y$, implying that $k \neq m$, let

$$h = \{(j, g_{k+j})\,|\,j \in [m-k]\} \cup \{(m-k+1, \{y, x\})\},$$

and check that h is a proper path from t to x in $\lambda(Q, \{s, t\}) - \{\{s, t\}\}$ and $h_{ph} = \{y, x\}.$

Thus, since in either case, h is a proper path from t to x in
$\lambda(Q, \{s, t\}) - \{\{s, t\}\}$ and $h_{ph} = \{y, x\}$, Lemma 1 applies to give the following four conditions.

(5) $(t, s) \in K$ and $(y, x) \in K$
     implies
     $(c(Q, \{s, t\}))\,(y, x) = -1.$

(6) $(t, s) \in K$ and $(x, y) \in K$
     implies
     $(c(Q, \{s, t\}))\,(x, y) = 1.$

(7) $(s, t) \in K$ and $(y, x) \in K$
implies
$(c(Q, \{s, t\}))\, (y, x) = 1$.

(8) $(s, t) \in K$ and $(x, y) \in K$
implies
$(c(Q, \{s, t\}))\, (x, y) = -1$.

A comparison of Condition (1) and (8), Conditions (2) and (6), Conditions (3) and (7), and Conditions (4) and (5) completes the proof of the Lemma.

Lemma 4. *Let $T \in \mathcal{T}$. Let $(x, y) \in K$. Let $(s, t) \in K$.*
Then,
$$(A_T(\Phi^K_{(x,y)}))\, (s, t) \neq 0$$
if and only if
$$\{x, y\} \in S - T \quad \text{and} \quad \{s, t\} \in \lambda(T, \{x, y\}).$$

Proof: Suppose first that $(A_T(\Phi^K_{(x,y)}))\, (s, t) \neq 0$.

Now, if $\{x, y\} \in T$, then $A_T(\Phi^K_{(x,y)}) = \theta_1$, and $(A_T(\Phi^K_{(x,y)}))\, (s, t) = 0$, which is impossible. Thus, $\{x, y\} \in S - T$, and $A_T(\Phi^K_{(x,y)}) = c(T, \{x, y\})$. Hence, $(A_T(\Phi^K_{(x,y)}))\, (s, t) = (c(T, \{x, y\}))\, (s, t)$, and by the definition of $c(T, \{x, y\})$ in 7.5, the fact that $(c(T, \{x, y\}))\, (s, t) \neq 0$ implies that $\{s, t\} \in \lambda(T, \{x, y\})$.

In the other direction, suppose that $\{x, y\} \in S - T$ and $\{s, t\} \in \lambda(T, \{x, y\})$.

Then, $A_T(\Phi^K_{(x,y)}) = c(T, \{x, y\})$, and by the definition of $c(T, \{x, y\})$ in 7.5, $(A_T(\Phi^K_{(x,y)}))\, (s, t) = (c(T, \{x, y\}))\, (s, t) \neq 0$.

Lemma 5. *Let $T \in \mathcal{T}$. Let $(x, y) \in K$. Let $(s, t) \in K$. Suppose that $\{x, y\} \neq \{s, t\}$. Suppose that $(A_T(\Phi^K_{(x,y)}))\, (s, t) \neq 0$.*
Then,
$$(A_T(\Phi^K_{(s,t)}))\, (x, y) = 0.$$

Proof: By Lemma 4, $\{x, y\} \in S - T$ and $\{s, t\} \in \lambda(T, \{x, y\}) \subset T \cup \{\{x, y\}\}$. Thus, since $\{s, t\} \neq \{x, y\}$, $\{s, t\} \in T$ and $A_T(\Phi^K_{(s,t)}) = \theta_1$, implying that $(A_T(\Phi^K_{(s,t)}))\, (x, y) = 0$.

Theorem 2. $Z \circ F = (Z \circ F)^t$.

Proof: We apply Theorem 1 of 4.7. Let $(x, y) \in K$ and let $(s, t) \in K$. It suffices to show that

$$\langle (Z \circ F)\, (\Phi^K_{(x,y)}), \Phi^K_{(s,t)} \rangle = \langle (Z \circ F)\, (\Phi^K_{(s,t)}), \Phi^K_{(x,y)} \rangle.$$

Now, if $\{x, y\} = \{s, t\}$, then $(x, y) = f(\{x, y\}) = f(\{s, t\}) = (s, t)$ and we are through. Thus, we may assume that $\{x, y\} \neq \{s, t\}$.

Now,

$$\langle (Z \circ F)\,(\Phi^K_{(x,y)})\,,\,\Phi^K_{(s,t)}\rangle = \langle Z(F(\Phi^K_{(x,y)}))\,,\,\Phi^K_{(s,t)}\rangle$$

$$= \langle Z((\sum_{T \in \mathcal{T}} W(T)\,A_T)\,(\Phi^K_{(x,y)}))\,,\,\Phi^K_{(s,t)}\rangle$$

$$= \langle Z(\sum_{T \in \mathcal{T}} W(T)\,A_T(\Phi^K_{(x,y)}))\,,\,\Phi^K_{(s,t)}\rangle$$

$$= \langle \sum_{T \in \mathcal{T}} W(T)\,Z(A_T(\Phi^K_{(x,y)}))\,,\,\Phi^K_{(s,t)}\rangle$$

$$= \sum_{T \in \mathcal{T}} W(T)\,\langle Z(A_T(\Phi^K_{(x,y)}))\,,\,\Phi^K_{(s,t)}\rangle.$$

Similarly,

$$\langle (Z \circ F)\,(\Phi^K_{(s,t)})\,,\,\Phi^K_{(x,y)}\rangle = \sum_{T \in \mathcal{T}} W(T)\,\langle Z(A_T(\Phi^K_{(s,t)}))\,,\,\Phi^K_{(x,y)}\rangle.$$

Thus, to complete the proof we must show that

$$\sum_{T \in \mathcal{T}} W(T)\,\langle Z(A_T(\Phi^K_{(x,y)}))\,,\,\Phi^K_{(s,t)}\rangle = \sum_{T \in \mathcal{T}} W(T)\,\langle Z(A_T(\Phi^K_{(s,t)}))\,,\,\Phi^K_{(x,y)}\rangle.$$

To show these two sums equal we invoke the Assumption concerning summation introduced in 5.5.
In particular, it suffices to exhibit a univalent function h with dmn $h = \mathcal{T}$ and rng $h = \mathcal{T}$, such that for each $T \in \mathcal{T}$,

$$W(T)\,\langle Z(A_T(\Phi^K_{(x,y)}))\,,\,\Phi^K_{(s,t)}\rangle = W(h(T))\,\langle Z(A_{h(T)}(\Phi^K_{(s,t)}))\,,\,\Phi^K_{(x,y)}\rangle$$

We proceed to specify this function h.

First note Lemma 5 which states that if $(A_T(\Phi^K_{(x,y)}))\,(s,t) \neq 0$, then $(A_T(\Phi^K_{(s,t)}))\,(x,y) = 0$, and also states that if $(A_T(\Phi^K_{(s,t)}))\,(x,y) \neq 0$, then $(A_T(\Phi^K_{(x,y)}))\,(s,t) = 0$.

Thus, to specify $h(T)$ for each $T \in \mathcal{T}$, it suffices first to specify $h(T)$ for $T \in \mathcal{T}$ such that $(A_T(\Phi^K_{(x,y)}))\,(s,t) \neq 0$, and then to specify $h(T)$ for $T \in \mathcal{T}$ such that $(A_T(\Phi^K_{(s,t)}))\,(x,y) \neq 0$, and finally to specify $h(T)$ for $T \in \mathcal{T}$ such that $(A_T(\Phi^K_{(x,y)}))\,(s,t) = 0$ and $(A_T(\Phi^K_{(s,t)}))\,(x,y) = 0$.

In particular, we let h be that function with domain $\mathcal{T}$ such that for each $T \in \mathcal{T}$,

$$h(T) = (T \cup \{\{x,y\}\}) - \{\{s,t\}\}, \text{ if } (A_T(\Phi^K_{(x,y)}))\,(s,t) \neq 0,$$

$$h(T) = (T \cup \{\{s,t\}\}) - \{\{x,y\}\}, \text{ if } (A_T(\Phi^K_{(s,t)}))\,(x,y) \neq 0,$$

$$h(T) = T, \text{ if } (A_T(\Phi^K_{(x,y)}))\,(s,t) = 0 = (A_T(\Phi^K_{(s,t)}))\,(x,y).$$

In the remainder of the proof which is divided into four parts we show that rng $h = \mathcal{T}$, h is univalent, and for each $T \in \mathcal{T}$,

$$W(T)\,\langle Z(A_T(\Phi^K_{(x,y)}))\,,\,\Phi^K_{(s,t)}\rangle = W(h(T))\,\langle Z(A_{h(T)}(\Phi^K_{(s,t)}))\,,\,\Phi^K_{(x,y)}\rangle.$$

Part 1. rng $h \subset \mathcal{T}$.

Proof of Part 1: Let $T \in \mathcal{T}$.

If $(A_T(\Phi^K_{(x,y)}))\,(s,t) \neq 0$, then $h(T) = (T \cup \{\{x,y\}\}) - \{\{s,t\}\}$, and by Lemma 4, $\{x,y\} \in S - T$ and $\{s,t\} \in \lambda(T, \{x,y\})$, and by Lemma 1, (2), $(T \cup \{\{x,y\}\}) - \{\{s,t\}\} \in \mathcal{T}$.

If $(A_T(\Phi^K_{(s,t)}))\,(x,y) \neq 0$, then $h(T) = (T \cup \{\{s,t\}\}) - \{\{x,y\}\}$, and by Lemma 4, $\{s,t\} \in S - T$ and $\{x,y\} \in \lambda(T, \{s,t\})$, and by Lemma 1, (2), $(T \cup \{\{s,t\}\}) - \{\{x,y\}\} \in \mathcal{T}$.

If $(A_T(\Phi^K_{(x,y)}))\,(s,t) = 0$ and $(A_T(\Phi^K_{(s,t)}))\,(x,y) = 0$, then $h(T) = T \in \mathcal{T}$.

Part 2. rng $h = \mathcal{T}$.

Proof of Part 2: Let $Q \in \mathcal{T}$. We shall exhibit $T \in \mathcal{T}$ such that $h(T) = Q$.

Three cases arise.

Case 1. $(A_Q(\Phi^K_{(x,y)}))\,(s,t) \neq 0$.

In this case, let $T = (Q \cup \{\{x,y\}\}) - \{\{s,t\}\}$.

By Lemma 4, $\{x,y\} \in S - Q$ and $\{s,t\} \in \lambda(Q, \{x,y\})$.
Thus, by Lemma 1, (2), $(Q \cup \{\{x,y\}\}) - \{\{s,t\}\} \in \mathcal{T}$, and $T \in \mathcal{T}$.
We shall show that $h(T) = Q$.

By Lemma 1, (3), $\lambda(Q, \{x,y\}) = \lambda(T, \{s,t\})$.
But $\{x,y\} \in \lambda(Q, \{x,y\})$.
Thus, $\{x,y\} \in \lambda(T, \{s,t\})$.
But $\{s,t\} \in S - T$.
Thus, by Lemma 4, $(A_T(\Phi^K_{(s,t)}))\,(x,y) \neq 0$.
By the definition of h,

$$h(T) = (T \cup \{\{s,t\}\}) - \{\{x,y\}\}$$
$$= (((Q \cup \{\{x,y\}\}) - \{\{s,t\}\}) \cup \{\{s,t\}\}) - \{\{x,y\}\}$$
$$= Q.$$

Case 2. $(A_Q(\Phi^K_{(s,t)}))\,(x,y) \neq 0$.

In this case, let $T = (Q \cup \{\{s,t\}\}) - \{\{x,y\}\}$.

By Lemma 4, $\{s,t\} \in S - Q$ and $(x,y) \in \lambda(Q, \{s,t\})$.
Thus, by Lemma 1, (2), $(Q \cup \{\{s,t\}\}) - \{\{x,y\}\} \in \mathcal{T}$, and $T \in \mathcal{T}$.
We shall show that $h(T) = Q$.

By Lemma 1, (3), $\lambda(Q, \{s,t\}) = \lambda(T, \{x,y\})$.
But $\{s,t\} \in \lambda(Q, \{s,t\})$.
Thus, $\{s,t\} \in \lambda(T, \{x,y\})$.
But $\{x,y\} \in S - T$.
Thus, by Lemma 4, $(A_T(\Phi^K_{(x,y)}))\,(s,t) \neq 0$.
By the definition of h,

$$h(T) = (T \cup \{\{x, y\}\}) - \{\{s, t\}\}$$
$$= (((Q \cup \{\{s, t\}\}) - \{\{x, y\}\}) \cup \{\{x, y\}\}) - \{\{s, t\}\}$$
$$= Q.$$

Case 3. $(A_Q(\Phi^K_{(x, y)}))\,(s, t) = 0 = (A_Q(\Phi^K_{(s, t)}))\,(x, y).$
In this case, let $T = Q$. Then, by the definition of h, $h(T) = h(Q) = Q$.

Part 3. h is univalent.

Proof of Part 3: Let $T \in \mathscr{T}$ and let $Q \in \mathscr{T}$ such that $h(T) = h(Q)$. We shall show that $T = Q$.

Because T and Q are arbitrary, it suffices to consider the following six cases.

Case 1. $(A_T(\Phi^K_{(x, y)}))\,(s, t) \neq 0$ and $(A_Q(\Phi^K_{(x, y)}))\,(s, t) \neq 0.$
In this case, since $h(T) = h(Q)$,

$$(T \cup \{\{x, y\}\}) - \{\{s, t\}\} = (Q \cup \{\{x, y\}\}) - \{\{s, t\}\}.$$

But by Lemma 4, $\{x, y\} \in S - T.$
Also, by Lemma 4, $\{s, t\} \in \lambda(T, \{x, y\}) \subset T \cup \{\{x, y\}\}$, and since $\{x, y\} \neq \{s, t\}$, $\{s, t\} \in T.$
Similarly, by Lemma 4, $\{x, y\} \in S - Q$ and $\{s, t\} \in Q.$
These facts imply that

$$T = (((T \cup \{\{x, y\}\}) - \{\{s, t\}\}) \cup \{\{s, t\}\}) - \{\{x, y\}\}$$
$$= (((Q \cup \{\{x, y\}\}) - \{\{s, t\}\}) \cup \{\{s, t\}\}) - \{\{x, y\}\}$$
$$= Q.$$

Case 2. $(A_T(\Phi^K_{(x, y)}))\,(s, t) \neq 0$ and $(A_Q(\Phi^K_{(s, t)}))\,(x, y) \neq 0.$
In this case, since $h(T) = h(Q)$,

$$(T \cup \{\{x, y\}\}) - \{\{s, t\}\} = (Q \cup \{\{s, t\}\}) - \{\{x, y\}\}.$$

But this is impossible, since $\{s, t\} \in (Q \cup \{\{s, t\}\}) - \{\{x, y\}\}$, but $\{s, t\} \notin (T \cup \{\{x, y\}\}) - \{\{s, t\}\}.$

Case 3. $(A_T(\Phi^K_{(x, y)}))\,(s, t) \neq 0$ and
$(A_Q(\Phi^K_{(x, y)}))\,(s, t) = 0 = (A_Q(\Phi^K_{(s, t)}))\,(x, y).$
In this case, since $h(T) = h(Q)$,

$$(T \cup \{\{x, y\}\}) - \{\{s, t\}\} = Q.$$

We shall show that this is impossible.

By Lemma 4, $\{s, t\} \in Q$ or $\{x, y\} \notin \lambda(Q, \{s, t\}).$
If $\{s, t\} \in Q$, then $\{s, t\} \in (T \cup \{\{x, y\}\}) - \{\{s, t\}\}$, which is impossible. Thus, $\{x, y\} \notin \lambda(Q, \{s, t\}).$

But again by Lemma 4, $\{x, y\} \in S - T$ and $\{s, t\} \in \lambda(T, \{x, y\})$.
By Lemma 1, (3), $\lambda(T, \{x, y\}) = \lambda(Q, \{s, t\})$.
But we have shown above that $\{x, y\} \notin \lambda(Q, \{s, t\})$.
Thus, $\{x, y\} \notin \lambda(T, \{x, y\})$, which is impossible.

Case 4. $(A_T(\Phi^K_{(s,t)}))\,(x, y) \neq 0$ and $(A_Q(\Phi^K_{(s,t)}))\,(x, y) \neq 0$.
This case is essentially the same as Case 1, except that the roles of $\{x, y\}$ and $\{s, t\}$ are interchanged.

Case 5. $(A_T(\Phi^K_{(s,t)}))\,(x, y) \neq 0$ and
$(A_Q(\Phi^K_{(x,y)}))\,(s, t) = 0 = (A_Q(\Phi^K_{(s,t)}))\,(x, y)$.
This case is essentially the same as Case 3, except that the roles of $\{x, y\}$ and $\{s, t\}$ are interchanged.

Case 6. $(A_T(\Phi^K_{(x,y)}))\,(s, t) = 0 = (A_T(\Phi^K_{(s,t)}))\,(x, y)$
and $(A_Q(\Phi^K_{(x,y)}))\,(s, t) = 0 = (A_Q(\Phi^K_{(s,t)}))\,(x, y)$.
In this case, $T = h(T) = h(Q) = Q$.

Part 4. $T \in \mathscr{T}$
implies

$$W(T)\,\langle Z(A_T(\Phi^K_{(x,y)})),\, \Phi^K_{(s,t)}\rangle = W(h(T))\,\langle Z(A_{h(T)}(\Phi^K_{(s,t)})),\, \Phi^K_{(x,y)}\rangle.$$

Proof of Part 4: Let $T \in \mathscr{T}$.
Note that

$$
\begin{aligned}
\langle Z(A_T(\Phi^K_{(x,y)})),\, \Phi^K_{(s,t)}\rangle &= \langle Z\big(\sum_{(u,v)\in K} (A_T(\Phi^K_{(x,y)}))\,(u, v)\,\Phi^K_{(u,v)}\big),\, \Phi^K_{(s,t)}\rangle \\
&= \langle \sum_{(u,v)\in K} (A_T(\Phi^K_{(x,y)}))\,(u, v)\,Z(\Phi^K_{(u,v)}),\, \Phi^K_{(s,t)}\rangle \\
&= \langle \sum_{(u,v)\in K} (A_T(\Phi^K_{(x,y)}))\,(u, v)\,r(\{u, v\})\,\Phi^K_{(u,v)},\, \Phi^K_{(s,t)}\rangle \\
&= \sum_{(u,v)\in K} (A_T(\Phi^K_{(x,y)}))\,(u, v)\,r(\{u, v\})\,\langle \Phi^K_{(u,v)},\, \Phi^K_{(s,t)}\rangle \\
&= (A_T(\Phi^K_{(x,y)}))\,(s, t)\,r(\{s, t\}).
\end{aligned}
$$

Similarly,

$$\langle Z(A_{h(T)}(\Phi^K_{(s,t)})),\, \Phi^K_{(x,y)}\rangle = (A_{h(T)}(\Phi^K_{(s,t)}))\,(x, y)\,r(\{x, y\}).$$

Thus, it suffices to show that

$$W(T)\,r(\{s, t\})\,(A_T(\Phi^K_{(x,y)}))\,(s, t) = W(h(T))\,r(\{x, y\})\,(A_{h(T)}(\Phi^K_{(s,t)}))\,(x, y).$$

Three cases arise.

Case 1. $(A_T(\Phi^K_{(x,y)}))\,(s, t) \neq 0$.
In this case, by Lemma 4, $\{x, y\} \in S - T$ and $A_T(\Phi^K_{(x,y)}) = c(T, \{x, y\})$.

Thus,

$$W(T)\, r(\{s,\,t\})\, (A_T(\Phi^K_{(x,\,y)}))\,(s,\,t)$$

$$= r(\{s,\,t\})\prod_{\{u,\,v\}\,\in\,S-T} r(\{u,\,v\})\,(c(T,\,\{x,\,y\}))\,(s,\,t)$$

$$= \prod_{\{u,\,v\}\,\in\,(S-T)\,\cup\,\{\{s,\,t\}\}} r(\{u,\,v\})\,(c(T,\,\{x,\,y\}))\,(s,\,t).$$

On the other hand,

$$h(T) = (T\cup\{\{x,\,y\}\}) - \{\{s,\,t\}\}.$$

Thus, $\{s,\,t\}\in S - h(T)$, and $A_{h(T)}(\Phi^K_{(s,\,t)}) = c(h(T),\,\{s,\,t\})$.
Hence,

$$W(h(T))\, r(\{x,\,y\})\, (A_{h(T)}(\Phi^K_{(s,\,t)}))\,(x,\,y)$$

$$= r(\{x,\,y\})\prod_{\{u,\,v\}\,\in\,S-h(T)} r(\{u,\,v\})\,(c(h(T),\,\{s,\,t\}))\,(x,\,y)$$

$$= \prod_{\{u,\,v\}\,\in\,(S-h(T))\,\cup\,\{\{x,\,y\}\}} r(\{u,\,v\})\,(c(h(T),\,\{s,\,t\}))\,(x,\,y).$$

But by Lemma 4, $\{x,\,y\}\in S - T$ and $\{s,\,t\}\in\lambda(T,\,\{x,\,y\})$.
By Lemma 3,

$$(c(T,\,\{x,\,y\}))\,(s,\,t) = (c(h(T),\,\{s,\,t\}))\,(x,\,y).$$

Thus, to complete this case, it suffices to show that

$$(S - T)\cup\{\{s,\,t\}\} = (S - h(T))\cup\{\{x,\,y\}\},$$

which is true because $\{x,\,y\}\in S - T$.

    Case 2. $(A_T(\Phi^K_{(s,\,t)}))\,(x,\,y)\neq 0$.

    In this case, by Lemma 5, $(A_T(\Phi^K_{(x,\,y)}))\,(s,\,t) = 0$, and

$$W(T)\, r(\{s,\,t\})\, (A_T(\Phi^K_{(x,\,y)}))\,(s,\,t) = 0.$$

Thus, it suffices to show that $(A_{h(T)}(\Phi^K_{(s,\,t)}))\,(x,\,y) = 0$.
But,

$$\{s,\,t\}\in (T\cup\{\{s,\,t\}\}) - \{\{x,\,y\}\} = h(T).$$

Hence, $A_{h(T)}(\Phi^K_{(s,\,t)}) = \theta_1$ and $(A_{h(T)}(\Phi^K_{(s,\,t)}))\,(x,\,y) = 0$.

    Case 3. $(A_T(\Phi^K_{(x,\,y)}))\,(s,\,t) = 0 = (A_T(\Phi^K_{(s,\,t)}))\,(x,\,y)$.

    In this case,

$$W(T)\, r(\{s,\,t\})\, (A_T(\Phi^K_{(x,\,y)}))\,(s,\,t) = 0,$$

and it suffices to show that $(A_{h(T)}(\Phi^K_{(s,\,t)}))\,(x,\,y) = 0$, which is true
since $h(T) = T$.

The proof is complete.

8.4 The Sum of Tree Chord Products

The sum of all tree chord products, $\sum\limits_{T \in \mathscr{T}} W(T)$, will occur sufficiently often for us to endow it with a special symbol. In particular, we shall let D be equal to the sum of the tree chord products. Formally,

$$D = \sum_{T \in \mathscr{T}} W(T).$$

Note from the definition of W in 8.3 that D is a positive real number.

The chord map F, applied to any cycle, merely yields the number D multiplied by this cycle. We exhibit this fact as the following Theorem.

Theorem. *Let $u \in$ kernel $\varDelta$.*
Then,
$F(u) = Du.$

Proof: By Theorem 2 of 8.2,

$$F(u) = \left(\sum_{T \in \mathscr{T}} W(T)\, A_T \right)(u)$$

$$= \sum_{T \in \mathscr{T}} W(T)\, A_T(u)$$

$$= \sum_{T \in \mathscr{T}} W(T)\, u$$

$$= \left(\sum_{T \in \mathscr{T}} W(T) \right) u$$

$$= Du.$$

As an immediate Corollary to the preceding Theorem we exhibit the fact that the restriction of the composition function $Z \circ F$ to the cycle space, kernel $\varDelta$, is equal to the restriction of the diagonal linear map DZ to the cycle space, kernel $\varDelta$.

Corollary. $(Z \circ F)|\text{kernel } \varDelta = (DZ)|\text{kernel } \varDelta.$
Proof: Let $u \in$ kernel $\varDelta$.
It suffices to show that $(Z \circ F)(u) = DZ(u).$
But by the preceding Theorem,

$$(Z \circ F)(u) = Z(F(u)) = Z(Du) = DZ(u).$$

8.5 The Current Chain with Voltage Sources

We are now in a position to specify the current chain $i \in \mathscr{L}(K)$ when our resistive network is energized exclusively with voltage sources. Recall, however, that one representation of the current chain has already been exhibited in Theorem 1 of 7.3, establishing the existence of the current chain and voltage chain satisfying Ohm's Law, Kirchhoff's Voltage Law, and Kirchhoff's Current Law. In this Chapter we seek another representation for the current chain $i \in \mathscr{L}(K)$ which will allow us to calculate easily the branch current $i(x, y)$ for each directed branch (x, y) of K.

To make matters precise, let the voltage source chain be $e \in \mathscr{L}(K)$. By Theorem 1 of 7.3, we are assured of the existence of a current chain $i \in \mathscr{L}(K)$ such that i is a cycle and $Z(i) - e$ is a coboundary. Thus,

$$i \in \text{kernel } \varDelta,$$
$$Z(i) - e \in \text{rng } \delta.$$

Before exhibiting the formula for the branch current $i(x, y)$ for each directed branch (x, y) of K, we need a preliminary Theorem 1 providing us with the value of the diagonal linear map DZ evaluated at i.

Theorem 1. *Let $e \in \mathscr{L}(K)$. Let $i \in \text{kernel } \varDelta$. Suppose that $Z(i) - e \in \text{rng } \delta$. Then,*
$$DZ(i) = \sum_{(s,\,t)\in K} e(s, t)\, F^t(\varPhi^K_{(s,\,t)}).$$

Proof: By Theorem 1 of 8.3, $F^t(Z(i) - e) = \theta_1$.
Thus, $F^t(Z(i)) - F^t(e) = F^t(Z(i) - e) = \theta_1$, implying that $F^t(Z(i)) = F^t(e)$.
But by Theorem 8 of 4.7, $Z^t = Z$.
Thus, by Theorem 7 of 4.7,

$$F^t(e) = F^t(Z(i)) = (F^t \circ Z)(i) = (F^t \circ Z^t)(i) = (Z \circ F)^t(i).$$

Hence, by Theorem 2 of 8.3 and the Corollary of 8.4,

$$F^t(e) = (Z \circ F)^t(i) = (Z \circ F)(i) = DZ(i).$$

But also,

$$F^t(e) = F^t\left(\sum_{(s,\,t)\in K} e(s, t)\, \varPhi^K_{(s,\,t)} \right) = \sum_{(s,\,t)\in K} e(s, t)\, F^t(\varPhi^K_{(s,\,t)}).$$

Thus, the proof is complete.

As the following Theorem 2 we can now specify a usable formula for the branch current $i(x, y)$ for each directed branch (x, y) of K. The formula exhibited in Theorem 2 is frequently called Kirchhoff's Third Law.

Theorem 2. *Let $e \in \mathscr{L}(K)$. Let $i \in \text{kernel } \varDelta$. Suppose that $Z(i) - e \in \text{rng } \delta$. Let $(x, y) \in K$. Then,*

$$i(x, y) = D^{-1}\big(r(\{x, y\})\big)^{-1} \sum_{(s,\,t)\in K} e(s, t) \sum_{T \in \mathscr{T}} W(T)\, (A_T(\varPhi^K_{(x,\,y)}))\,(s, t).$$

Proof: Note that

$$DZ(i) = DZ\Big(\sum_{(s,\,t)\,\in\,K} i(s,\,t)\,\Phi^K_{(s,\,t)}\Big)$$

$$= D \sum_{(s,\,t)\,\in\,K} i(s,\,t)\, Z(\Phi^K_{(s,\,t)})$$

$$= D \sum_{(s,\,t)\,\in\,K} i(s,\,t)\, r(\{s,\,t\})\, \Phi^K_{(s,\,t)}.$$

Thus,

$$\langle DZ(i),\, \Phi^K_{(x,\,y)}\rangle = \langle D \sum_{(s,\,t)\,\in\,K} i(s,\,t)\, r(\{s,\,t\})\, \Phi^K_{(s,\,t)},\, \Phi^K_{(x,\,y)}\rangle$$

$$= D \sum_{(s,\,t)\,\in\,K} i(s,\,t)\, r(\{s,\,t\})\, \langle \Phi^K_{(s,\,t)},\, \Phi^K_{(x,\,y)}\rangle$$

$$= D\,r(\{x,\,y\})\, i(x,\,y).$$

Thus,

$$i(x,\,y) = D^{-1}(r(\{x,\,y\}))^{-1}\langle DZ(i),\, \Phi^K_{(x,\,y)}\rangle,$$

and to evaluate $\langle DZ(i),\, \Phi^K_{(x,\,y)}\rangle$ we use Theorem 1.

Noting Theorem 2 of 4.7,

$$\langle DZ(i),\, \Phi^K_{(x,\,y)}\rangle = \langle \sum_{(s,\,t)\,\in\,K} e(s,\,t)\, F^t(\Phi^K_{(s,\,t)}),\, \Phi^K_{(x,\,y)}\rangle$$

$$= \sum_{(s,\,t)\,\in\,K} e(s,\,t)\, \langle F^t(\Phi^K_{(s,\,t)}),\, \Phi^K_{(x,\,y)}\rangle$$

$$= \sum_{(s,\,t)\,\in\,K} e(s,\,t)\, \langle F(\Phi^K_{(x,\,y)}),\, \Phi^K_{(s,\,t)}\rangle$$

$$= \sum_{(s,\,t)\,\in\,K} e(s,\,t)\, (F(\Phi^K_{(x,\,y)}))\,(s,\,t)$$

$$= \sum_{(s,\,t)\,\in\,K} e(s,\,t)\, \Big(\big(\sum_{T\,\in\,\mathscr{T}} W(T)\, A_T\big)(\Phi^K_{(x,\,y)})\Big)\,(s,\,t)$$

$$= \sum_{(s,\,t)\,\in\,K} e(s,\,t)\, \Big(\sum_{T\,\in\,\mathscr{T}} W(T)\, A_T(\Phi^K_{(x,\,y)})\Big)\,(s,\,t)$$

$$= \sum_{(s,\,t)\,\in\,K} e(s,\,t)\, \sum_{T\,\in\,\mathscr{T}} W(T)\, (A_T(\Phi^K_{(x,\,y)}))\,(s,\,t).$$

Hence, since

$$i(x,\,y) = D^{-1}(r(\{x,\,y\}))^{-1}\langle DZ(i),\, \Phi^K_{(x,\,y)}\rangle,$$

we obtain the desired result,

$$i(x,\,y) = D^{-1}(r(\{x,\,y\}))^{-1} \sum_{(s,\,t)\,\in\,K} e(s,\,t)\, \sum_{T\,\in\,\mathscr{T}} W(T)\, (A_T(\Phi^K_{(x,\,y)}))\,(s,\,t).$$

The result of the preceding Theorem 2 does not appear, at first glance, to offer an easily calculated branch current $i(x,\,y)$ for each directed branch (x, y) of K. However, to analyze this formula we consider a special case of the voltage source chain $e \in \mathscr{L}(K)$ when e is such that it assigns a zero source voltage to all directed branches of K except at most one. Such a source chain can be written as an element $w\,\Phi^K_{(s, t)}$ of $\mathscr{L}(K)$ where w is a real number and (s, t) is a directed branch of K.

In the following Corollary we reduce the formula exhibited in Theorem 2 to the special case of this source chain $w\,\Phi^K_{(s, t)}$. Since the Corollary follows trivially from the preceding Theorem 2, we omit the proof.

Corollary. *Let* $w \in R$. *Let* $(s, t) \in K$. *Let* $i \in \mathrm{kernel}\ \Delta$. *Suppose that* $Z(i) - w\,\Phi^K_{(s, t)} \in \mathrm{rng}\ \delta$. *Let* $(x, y) \in K$.
Then,

$$i(x, y) = D^{-1}\left(r(\{x, y\})\right)^{-1} w \sum_{T \in \mathscr{T}} W(T)\, \left(A_T(\Phi^K_{(x, y)})\right)(s, t).$$

Now, returning again to the general source chain $e \in \mathscr{L}(K)$, we can write,

$$e = \sum_{(s, t)} e(s, t)\, \Phi^K_{(s, t)},$$

and thus, e is a sum of the special type of source chains considered in the Corollary, each of which assigns a non-zero source voltage to at most one directed branch of K. Observe that the branch current $i(x, y)$ resulting from this general source chain $e \in \mathscr{L}(K)$ is the sum of the individual branch currents, each resulting from consideration of the special type of source chains used in the Corollary. Thus, in loose language, we can state that the effect of sources upon the branch current $i(x, y)$ is additive, and to investigate the formula for the branch current we can concentrate upon the representation exhibited in the Corollary, in the case of the special source chain which assigns a non-zero source voltage to at most one directed branch of $\mathscr{L}(K)$.

Now, those trees $T \in \mathscr{T}$ which contribute a non-zero term to the formula for $i(x, y)$ exhibited in the Corollary are precisely those trees $T \in \mathscr{T}$ such that

$$\left(A_T(\Phi^K_{(x, y)})\right)(s, t) \neq 0.$$

Thus, Lemma 4 of 8.3 shows that we need only consider those trees $T \in \mathscr{T}$ such that $\{x, y\} \in S - T$ and $\{s, t\} \in \lambda(T, \{x, y\})$. This simple observation reduces tremendously the number of trees to be considered, for we first consider only those trees $T \in \mathscr{T}$ such that $\{x, y\}$ is not a branch of T, and then reduce this set of trees to its subset consisting of those trees such that $\{s, t\}$ is a branch of the unique loop formed by adjoining $\{x, y\}$ to the tree. This reduction in the number of trees to be considered makes the formula exhibited in the Corollary easily computable.

8.6 The Coboundary Map

The formulas of the preceding Paragraph provide us with information about the branch currents when our resistive network is energized exclusively by voltage sources. We now consider the dual situation and investigate the branch voltages when the resistive network is energized exclusively by current sources. The development for the calculation of the branch voltages in the presence of current sources will be consistently dual to the preceding development in which we calculated the branch currents in the presence of voltage sources. We shall call attention to this duality as we proceed.

For each $T \in \mathscr{T}$ we now define a coboundary map B_T which is dual to the cycle map A_T. If $T \in \mathscr{T}$ the coboundary map B_T will be a linear map of $\mathscr{L}(K)$ into $\mathscr{L}(K)$ which maps $\Phi^K_{(s,\,t)}$ into θ_1 when $(s, t) \in K$ such that $\{s, t\} \in S - T$, and which maps $\Phi^K_{(s,\,t)}$ into the coboundary $- d(T, \{s, t\})$ when $(s, t) \in K$ such that $\{s, t\} \in T$. Formally,

> For each $T \in \mathscr{T}$,
> B_T is the unique linear map of $\mathscr{L}(K)$ into $\mathscr{L}(K)$
> such that for each $(s, t) \in K$,
>
> (1) $B_T(\Phi^K_{(s,\,t)}) = \theta_1$, if $\{s, t\} \in S - T$.
> (2) $B_T(\Phi^K_{(s,\,t)}) = - d(T, \{s, t\})$, if $\{s, t\} \in T$.

We now investigate properties of the coboundary map B_T when $T \in \mathscr{T}$.

Suppose that $T \in \mathscr{T}$ and $(s, t) \in K$ such that $\{s, t\} \in T$. As our first property of the coboundary map B_T, we show in the following Theorem 1 that $d(T, \{s, t\})$ is invariant under the map B_T. The following Theorem 1 is the dual of Theorem 1 of 8.2. Before proceeding with Theorem 1, however, we need a preliminary Lemma.

Lemma. *Let $T \in \mathscr{T}$. Let $\{s, t\} \in T$. Suppose that $(s, t) \in K$. Then,*
$$(d(T, \{s, t\})) (s, t) = - 1.$$

Proof: Note that $\{s, t\} \in \mu(T, \{s, t\})$, and refer to Condition (3) in the definition of $d(T, \{s, t\})$ in 7.6.

It is immediate that $s \in \operatorname{star}^V_{T - \{\{s,\,t\}\}}(s)$.
Thus, it suffices to show that $t \notin \operatorname{star}^V_{T - \{\{s,\,t\}\}}(s)$.

Suppose that $t \in \operatorname{star}^V_{T - \{\{s,\,t\}\}}(s)$.
Then, there exists g such that g is a proper path from s to t in $T - \{\{s, t\}\}$. But it is clear that $\operatorname{rng} g \cup \{\{s, t\}\}$ is a loop and $\operatorname{rng} g \cup \{\{s, t\}\} \subset (T - \{\{s, t\}\}) \cup \{\{s, t\}\} = T$, which contradicts the fact that T is a tree.

Theorem 1. *Let $T \in \mathscr{T}$. Let $(s, t) \in K$. Suppose that $\{s, t\} \in T$. Then,*

$$B_T(d(T, \{s, t\})) = d(T, \{s, t\}).$$

Proof: Note that $d(T, \{s, t\}) = \sum_{(x, y) \in K} (d(T, \{s, t\}))\,(x, y)\,\Phi^K_{(x, y)}.$

Thus,

$$B_T(d(T, \{s, t\})) = B_T\Big(\sum_{(x, y) \in K} (d(T, \{s, t\}))\,(x, y)\,\Phi^K_{(x, y)}\Big)$$
$$= \sum_{(x, y) \in K} (d(T, \{s, t\}))\,(x, y)\,B_T(\Phi^K_{(x, y)}).$$

Now, if $(x, y) \in K$ such that $\{x, y\} \in S - T$, then $B_T(\Phi^K_{(x, y)}) = \theta_1$. Thus,

$$B_T(d(T, \{s, t\})) = \sum_{(x, y) \in K} (d(T, \{s, t\}))\,(x, y)\,B_T(\Phi^K_{(x, y)})$$
$$= \sum_{(x, y) \in \mathrm{rng}\,(f|T)} (d(T, \{s, t\}))\,(x, y)\,B_T(\Phi^K_{(x, y)}).$$

Now, suppose that $(x, y) \in \mathrm{rng}\,(f|T)$ with $(s, t) \neq (x, y)$ and consider $(d(T, \{s, t\}))\,(x, y)$. Since $\mu(T, \{s, t\}) \cap T = \{\{s, t\}\}$, and $\{x, y\} \in T$, we have $\{x, y\} \notin \mu(T, \{s, t\})$, and by Condition (2) of the definition of $d(T, \{s, t\})$ in 7.6, $(d(T, \{s, t\}))\,(x, y) = 0$. Thus,

$$B_T(d(T, \{s, t\})) = \sum_{(x, y) \in \mathrm{rng}\,(f|T)} (d(T, \{s, t\}))\,(x, y)\,B_T(\Phi^K_{(x, y)})$$
$$= (d(T, \{s, t\}))\,(s, t)\,B_T(\Phi^K_{(s, t)}).$$

But by the preceding Lemma, $(d(T, \{s, t\}))\,(s, t) = -1$. Thus,

$$B_T(d(T, \{s, t\})) = (d(T, \{s, t\}))\,(s, t)\,B_T(\Phi^K_{(s, t)})$$
$$= -B_T(\Phi^K_{(s, t)})$$
$$= d(T, \{s, t\}).$$

Again, let $T \in \mathscr{T}$. Using Theorem 1 it is easy to show that each coboundary of $\mathrm{rng}\,\delta$ is invariant under the coboundary map B_T. This fact is exhibited in the following Theorem 2 which is the dual of Theorem 2 of 8.2.

Theorem 2. *Let $T \in \mathscr{T}$. Let $z \in \mathrm{rng}\,\delta$.* *Then,*

$$B_T(z) = z.$$

Proof: By Theorem 2 of 7.6, $\{d(T, \{s, t\}) \mid \{s, t\} \in T\}$ is a base of $\mathrm{rng}\,\delta$. Thus, we can produce a function u such that $\mathrm{dmn}\,u = T$ and $\mathrm{rng}\,u \subset R$ and

$$z = \sum_{\{s,\,t\} \in T} u(\{s,t\}) \, d(T, \{s,t\}).$$

By Theorem 1,

$$\begin{aligned}
B_T(z) &= B_T\Big(\sum_{\{s,\,t\} \in T} u(\{s,t\}) \, d(T, \{s,t\}) \Big) \\
&= \sum_{\{s,\,t\} \in T} u(\{s,t\}) \, B_T(d(T, \{s,t\})) \\
&= \sum_{\{s,\,t\} \in T} u(\{s,t\}) \, d(T, \{s,t\}) \\
&= z.
\end{aligned}$$

Let $T \in \mathscr{T}$. From the definition of the coboundary map B_T it is apparent that the range of the coboundary map B_T is a subset of the coboundary space, range δ. From Theorem 2 it is easy to show that the range of the coboundary map is, in fact, equal to the coboundary space, range δ. We exhibit this fact as the following Theorem 3 which is the dual of Theorem 3 of 8.2.

Theorem 3. *Let $T \in \mathscr{T}$.*
Then,

$$\operatorname{rng} B_T = \operatorname{rng} \delta.$$

Proof: Let $w \in \operatorname{rng} B_T$.
Produce $u \in \mathscr{L}(K)$ such that $w = B_T(u)$.
Then,

$$\begin{aligned}
w &= B_T(u) \\
&= B_T\Big(\sum_{(s,\,t) \in K} u(s,t) \, \Phi^K_{(s,\,t)} \Big) \\
&= \sum_{(s,\,t) \in K} u(s,t) \, B_T(\Phi^K_{(s,\,t)}).
\end{aligned}$$

Let $(s,t) \in K$.
If $\{s,t\} \in S - T$, then $B_T(\Phi^K_{(s,\,t)}) = \theta_1 \in \operatorname{rng} \delta$, and $u(s,t) \, B_T(\Phi^K_{(s,\,t)}) \in \operatorname{rng} \delta$.
If $\{s,t\} \in T$, then $B_T(\Phi^K_{(s,\,t)}) = d(T, \{s,t\}) \in \operatorname{rng} \delta$, and
$u(s,t) \, B_T(\Phi^K_{(s,\,t)}) \in \operatorname{rng} \delta$.
Thus, $w \in \operatorname{rng} \delta$, and $\operatorname{rng} B_T \subset \operatorname{rng} \delta$.

On the other hand, let $z \in \operatorname{rng} \delta$.
By Theorem 2, $z = B_T(z) \in \operatorname{rng} B_T$.
Thus, $\operatorname{rng} \delta \subset \operatorname{rng} B_T$.

Again, let $T \in \mathscr{T}$. The last property of the coboundary map that we exhibit here concerns $(B_T)^t$, the transpose of the coboundary map, which we agree to denote more simply by B_T^t.

In the following Theorem 4, which is the dual of Theorem 4 of 8.2, we show that B_T^t, the transpose of the coboundary map B_T, sends each cycle into θ_1, the zero element of $\mathscr{L}(K)$.

Theorem 4. *Let $T \in \mathscr{T}$.*
Then,

rng $(B_T^t | \text{kernel } \Delta) = \{\theta_1\}$.

Proof: It suffices to show that rng $(B_T^t | \text{kernel } \Delta) \subset \{\theta_1\}$.

Let $z \in \text{kernel } \Delta$.

We shall show that $B_T^t(z) = \theta_1$.

But by Theorem 1 of 4.4, Theorem 2 of 4.5, and Theorem 2 of 4.7,

$$B_T^t(z) = \sum_{(s,t) \in K} (B_T^t(z)) \, (s,t) \, \Phi_{(s,t)}^K$$

$$= \sum_{(s,t) \in K} \langle B_T^t(z), \Phi_{(s,t)}^K \rangle \, \Phi_{(s,t)}^K$$

$$= \sum_{(s,t) \in K} \langle z, B_T(\Phi_{(s,t)}^K) \rangle \, \Phi_{(s,t)}^K.$$

Let $(s,t) \in K$.

By Theorem 3, $B_T(\Phi_{(s,t)}^K) \in \text{rng } \delta$.

By Theorem 1 of 5.9, $\langle z, B_T(\Phi_{(s,t)}^K) \rangle = 0$.

Thus, $B_T(z) = \theta_1$.

8.7 The Tree Branch Map

From all trees $T \in \mathscr{T}$, and the branches of such trees we shall manufacture a linear map of $\mathscr{L}(K)$ into $\mathscr{L}(K)$ and investigate its properties. This linear map to be manufactured will be called the tree branch map and will be denoted by G.

Before defining the tree branch map G, however, we must define a function U whose domain is equal to $\mathscr{T}$. For each $T \in \mathscr{T}$, we shall let $U(T)$ be the product of the reciprocale $(r(\{x,y\}))^{-1}$ of the values of the resistance function r, evaluated only for each branch $\{x,y\}$ of T. For each $T \in \mathscr{T}$ we call $U(T)$ the tree branch product. Formally,

U is the function with domain $\mathscr{T}$

such that for each $T \in \mathscr{T}$,

$$U(T) = \prod_{\{x,y\} \in T} (r(\{x,y\}))^{-1}.$$

The tree branch map G is now defined as the sum, for each $T \in \mathscr{T}$, of all linear maps $U(T) B_T$. Formally,

$$G = \sum_{T \in \mathscr{T}} U(T) B_T.$$

The first property of the tree branch map G, which we exhibit in the following Theorem 1, is the fact that the transpose G^t, of the tree branch map G, sends each cycle into θ_1, the zero element of $\mathscr{L}(K)$. Theorem 1 is the dual of Theorem 1 of 8.3.

Theorem 1. rng $(G^t | \text{kernel } \varDelta) = \{\theta_1\}$.

Proof: Clearly, G is a linear map of $\mathscr{L}(K)$ into $\mathscr{L}(K)$ and thus, G^t is a linear map of $\mathscr{L}(K)$ into $\mathscr{L}(K)$.

Hence, it suffices to show that rng $(G^t | \text{kernel } \varDelta) \subset \{\theta_1\}$.

Let $z \in \text{kernel } \varDelta$.

We shall show that $G^t(z) = \theta_1$.

But by Theorem 5 of 4.7 and Theorem 6 of 4.7,

$$G^t(z) = \Big(\sum_{T \in \mathscr{T}} U(T)\, B_T \Big)^t (z)$$

$$= \Big(\sum_{T \in \mathscr{T}} U(T)\, B_T^t \Big) (z)$$

$$= \sum_{T \in \mathscr{T}} U(T)\, B_T^t (z).$$

Let $T \in \mathscr{T}$.

By Theorem 4 of 8.6, $B_T^t(z) = \theta_1$, implying that $U(T)\, B_T^t(z) = \theta_1$. Thus, $G^t(z) = \theta_1$.

The next fact about the tree branch map G, which we shall exhibit in Theorem 2, is the assertion that the superposition function $Y \circ G$, is equal to its transpose, $(Y \circ G)^t$. This Theorem 2 of this Paragraph is the dual of Theorem 2 of 8.3. Thus, as in the case of Theorem 2 of 8.3, we expect some difficulty in the proof of Theorem 2 of this Paragraph.

As in the case of Theorem 2 of 8.3, we shall begin with a sucession of preliminary Lemmas, each describing a rather special situation. We introduce the Lemmas with no further comment. Then, at the conclusion of Lemmas we present Theorem 2, stating that $Y \circ G$ is equal to its transpose, $(Y \circ G)^t$, and show in the proof of Theorem 2 how the various Lemmas are needed.

Lemma 1. *Let* $T \in \mathscr{T}$. *Let* $\{s, t\} \in T$. *Let* $\{x, y\} \in S$. *Suppose that* $x \in \text{star}^V_{T-\{\{s,t\}\}}(s)$ *and* $y \notin \text{star}^V_{T-\{\{s,t\}\}}(s)$.

Then,

$x \notin \text{star}^V_{T-\{\{s,t\}\}}(t)$ *and* $y \in \text{star}^V_{T-\{\{s,t\}\}}(t)$.

Proof: By the Corollary of 2.16, either $x \in \text{star}^V_{T-\{\{s,t\}\}}(t)$ and $y \notin \text{star}^V_{T-\{\{s,t\}\}}(t)$, or $x \notin \text{star}^V_{T-\{\{s,t\}\}}(t)$ and $y \in \text{star}^V_{T-\{\{s,t\}\}}(t)$.

Thus, we assume that $x \in \text{star}^V_{T-\{\{s,t\}\}}(t)$ and $y \notin \text{star}^V_{T-\{\{s,t\}\}}(t)$. We shall show that this is impossible.

Since $x \in \text{star}^V_{T-\{\{s,t\}\}}(t)$, produce g such that g is a proper path from t to x in $T - \{\{s, t\}\}$.

From g it is easy to manufacture h such that h is a proper path from x to t in $T - \{\{s, t\}\}$.

Since $x \in \text{star}^V_{T-\{\{s,t\}\}}(s)$, produce u such that u is a proper path from s to x in $T - \{\{s, t\}\}$.

From u and h it is easy to manufacture w such that w is a proper path from s to t in $T - \{\{s, t\}\}$.

Then, $\text{rng } w \cup \{\{s, t\}\}$ is a loop and

$\text{rng } w \cup \{\{s, t\}\} \subset (T - \{\{s, t\}\}) \cup \{\{s, t\}\} = T$, which is impossible.

Lemma 2. *Let* $T \in \mathcal{F}$. *Let* $\{x, y\} \in S - T$. *Let* $\{s, t\} \in \lambda(T, \{x, y\})$. *Suppose that* $\{x, y\} \neq \{s, t\}$. *Let* $Q = (T \cup \{\{x, y\}\}) - \{\{s, t\}\}$. *Then,*

$$\left(d(T, \{s, t\})\right)\left(f(\{x, y\})\right) = \left(d(Q, \{x, y\})\right)\left(f(\{s, t\})\right).$$

Proof: The proof is divided into five parts.

Part 1. $\{x, y\} \in \mu(T, \{s, t\})$.

Proof of Part 1: Note that $\{s, t\} \in \lambda(T, \{x, y\}) \subset T \cup \{\{x, y\}\}$, but $\{s, t\} \neq \{x, y\}$.

Thus, $\{s, t\} \in T$, and we can consider $\mu(T, \{s, t\})$.

But $\{x, y\} \in S - T$ and $\{s, t\} \in \lambda(T, \{x, y\})$.

By Condition (3) of the definition of $\mu(T, \{s, t\})$ in 7.6, $\{x, y\} \in \mu(T, \{s, t\})$.

Part 2. $\{x, y\} \in Q$ and $Q \in \mathcal{F}$ and $\{s, t\} \in \mu(Q, \{x, y\})$.

Proof of Part 2: Note that $\{x, y\} \in Q$, and by Lemma 1, (2), of 8.3, $Q \in \mathcal{F}$. Thus, we can consider $\mu(Q, \{x, y\})$.

But by Lemma 1, (3) of 8.3, $\lambda(T, \{x, y\}) = \lambda(Q, \{s, t\})$.

Thus, since $\{x, y\} \in \lambda(T, \{x, y\})$, $\{x, y\} \in \lambda(Q, \{s, t\})$. But $\{s, t\} \in S - Q$.

By Condition (3) of the definition of $\mu(Q, \{x, y\})$ in 7.6, $\{s, t\} \in \mu(Q, \{x, y\})$.

Part 3. $x \in \text{star}^V_{T-\{\{s,t\}\}}(s)$ implies $s \in \text{star}^V_{Q-\{\{x,y\}\}}(x)$.

Proof of Part 3: Suppose that $x \in \text{star}^V_{T-\{\{s,t\}\}}(s)$.

Produce g such that g is a proper path from s to x in $T - \{\{s, t\}\}$.

Since $\{x, y\} \in S - T$, note that $\{x, y\} \notin \text{rng } g$.

But $\text{rng } g \subset T - \{\{s, t\}\} \subset (T \cup \{\{x, y\}\}) - \{\{s, t\}\} = Q$.

Since $\{x, y\} \notin \text{rng } g$, $\text{rng } g \subset Q - \{\{x, y\}\}$.

Thus, g is a proper path from s to x in $Q - \{\{x, y\}\}$, and it is easy to produce a proper path from x to s in $Q - \{\{x, y\}\}$, implying that $s \in \text{star}^V_{Q-\{\{x,y\}\}}(x)$.

Part 4. $y \in \text{star}^V_{T-\{\{s,t\}\}}(s)$ implies $s \in \text{star}^V_{Q-\{\{x,y\}\}}(y)$.

Proof of Part 4: The proof is essentially the same as the proof of Part 3, except that the roles of x and y are interchanged.

Part 5. $\left(d(T, \{s, t\})\right)\left(f(\{x, y\})\right) = \left(d(Q, \{x, y\})\right)\left(f(\{s, t\})\right)$.

Proof of Part 5. Four cases arise.

Case 1. $f(\{x, y\}) = (x, y)$ and $f(\{s, t\}) = (s, t)$.

By Part 1 and the Theorem of 2.16, either $x \in \operatorname{star}^V_{T-\{\{s,t\}\}}(s)$ and $y \notin \operatorname{star}^V_{T-\{\{s,t\}\}}(s)$, or $x \notin \operatorname{star}^V_{T-\{\{s,t\}\}}(s)$ and $y \in \operatorname{star}^V_{T-\{\{s,t\}\}}(s)$.

Suppose first that $x \in \operatorname{star}^V_{T-\{\{s,t\}\}}(s)$ and $y \notin \operatorname{star}^V_{T-\{\{s,t\}\}}(s)$. By Condition (3) of the definition of $d(T, \{s, t\})$ in 7.6,
$$\big(d(T, \{s, t\})\big)(x, y) = -1.$$
But by Part 2 and the Theorem of 2.16, either $s \in \operatorname{star}^V_{Q-\{\{x,y\}\}}(x)$ and $t \notin \operatorname{star}^V_{Q-\{\{x,y\}\}}(x)$, or $s \notin \operatorname{star}^V_{Q-\{\{x,y\}\}}(x)$ and $t \in \operatorname{star}^V_{Q-\{\{x,y\}\}}(x)$. But by Part 3, $s \in \operatorname{star}^V_{Q-\{\{x,y\}\}}(x)$, implying that $s \in \operatorname{star}^V_{Q-\{\{x,y\}\}}(x)$ and $t \notin \operatorname{star}^V_{Q-\{\{x,y\}\}}(x)$.

By Part 2 and Condition (3) of the definition of $d(Q, \{x, y\})$ in 7.6,
$$\big(d(Q, \{x, y\})\big)(s, t) = -1 = \big(d(T, \{s, t\})\big)(x, y).$$

On the other hand, suppose that $x \notin \operatorname{star}^V_{T-\{\{s,t\}\}}(s)$ and $y \in \operatorname{star}^V_{T-\{\{s,t\}\}}(s)$. By Condition (4) of the definition of $d(T, \{s, t\})$ in 7.6,
$$\big(d(T, \{s, t\})\big)(x, y) = 1.$$
But by Part 2 and the Theorem of 2.16, either $s \in \operatorname{star}^V_{Q-\{\{x,y\}\}}(y)$ and $t \notin \operatorname{star}^V_{Q-\{\{x,y\}\}}(y)$, or $s \notin \operatorname{star}^V_{Q-\{\{x,y\}\}}(y)$ and $t \in \operatorname{star}^V_{Q-\{\{x,y\}\}}(y)$. But by Part 4, $s \in \operatorname{star}^V_{Q-\{\{x,y\}\}}(y)$, implying that $s \in \operatorname{star}^V_{Q-\{\{x,y\}\}}(y)$ and $t \notin \operatorname{star}^V_{Q-\{\{x,y\}\}}(y)$.

By Part 2 and Lemma 1, $t \in \operatorname{star}^V_{Q-\{\{x,y\}\}}(x)$ and $s \notin \operatorname{star}^V_{Q-\{\{x,y\}\}}(x)$.

By Part 2 and Condition (4) of the definition of $d(Q, \{x, y\})$ in 7.6,
$$\big(d(Q, \{x, y\})\big)(s, t) = 1 = \big(d(T, \{s, t\})\big)(x, y).$$

Case 2. $f(\{x, y\}) = (x, y)$ and $f(\{s, t\}) = (t, s)$.

By Part 1 and the Theorem of 2.16, either $x \in \operatorname{star}^V_{T-\{\{s,t\}\}}(s)$ and $y \notin \operatorname{star}^V_{T-\{\{s,t\}\}}(s)$, or $x \notin \operatorname{star}^V_{T-\{\{s,t\}\}}(s)$ and $y \in \operatorname{star}^V_{T-\{\{s,t\}\}}(s)$.

Suppose first that $x \in \operatorname{star}^V_{T-\{\{s,t\}\}}(s)$ and $y \notin \operatorname{star}^V_{T-\{\{s,t\}\}}(s)$. By Lemma 1, $x \notin \operatorname{star}^V_{T-\{\{s,t\}\}}(t)$ and $y \in \operatorname{star}^V_{T-\{\{s,t\}\}}(t)$.

By Condition (6) of the definition of $d(T, \{s, t\})$ in 7.6,
$$\big(d(T, \{s, t\})\big)(x, y) = 1.$$
But by Part 2 and the Theorem of 2.16, either $s \in \operatorname{star}^V_{Q-\{\{x,y\}\}}(x)$ and $t \notin \operatorname{star}^V_{Q-\{\{x,y\}\}}(x)$, or $s \notin \operatorname{star}^V_{Q-\{\{x,y\}\}}(x)$ and $t \in \operatorname{star}^V_{Q-\{\{x,y\}\}}(x)$. But by Part 3, $s \in \operatorname{star}^V_{Q-\{\{x,y\}\}}(x)$, implying that $s \in \operatorname{star}^V_{Q-\{\{x,y\}\}}(x)$ and $t \notin \operatorname{star}^V_{Q-\{\{x,y\}\}}(x)$.

By Part 2 and Condition (4) of the definition of $d(Q, \{x, y\})$ in 7.6,
$$\big(d(Q, \{x, y\})\big)(t, s) = 1 = \big(d(T, \{s, t\})\big)(x, y).$$

On the other hand, suppose that $x \notin \operatorname{star}^V_{T-\{\{s,t\}\}}(s)$ and $y \in \operatorname{star}^V_{T-\{\{s,t\}\}}(s)$. By Lemma 1, $x \in \operatorname{star}^V_{T-\{\{s,t\}\}}(t)$ and $y \notin \operatorname{star}^V_{T-\{\{s,t\}\}}(t)$.

By Condition (5) of the definition of $d(T, \{s, t\})$ in 7.6,
$$\big(d(T, \{s, t\})\big)(x, y) = -1.$$
But by Part 2 and the Theorem of 2.16, either $s \in \operatorname{star}^V_{Q-\{\{x,y\}\}}(y)$ and $t \notin \operatorname{star}^V_{Q-\{\{x,y\}\}}(y)$, or $s \notin \operatorname{star}^V_{Q-\{\{x,y\}\}}(y)$ and $t \in \operatorname{star}^V_{Q-\{\{x,y\}\}}(y)$. But by Part 4, $s \in \operatorname{star}^V_{Q-\{\{x,y\}\}}(y)$, implying that $s \in \operatorname{star}^V_{Q-\{\{x,y\}\}}(y)$ and $t \notin \operatorname{star}^V_{Q-\{\{x,y\}\}}(y)$.

By Part 2 and Lemma 1, $s \notin \text{star}^V_{Q-\{\{x,y\}\}}(x)$ and $t \in \text{star}^V_{Q-\{\{x,y\}\}}(x)$.
By Part 2 and Condition (3) of the definition of $d(Q, \{x, y\})$ in 7.6,
$(d(Q, \{x, y\}))(t, s) = -1 = (d(T, \{s, t\}))(x, y)$.

Case 3. $f(\{x, y\}) = (y, x)$ and $f(\{s, t\}) = (s, t)$.

By Part 1 and the Theorem of 2.16, either $x \in \text{star}^V_{T-\{\{s,t\}\}}(s)$ and $y \notin \text{star}^V_{T-\{\{s,t\}\}}(s)$, or $x \notin \text{star}^V_{T-\{\{s,t\}\}}(s)$ and $y \in \text{star}^V_{T-\{\{s,t\}\}}(s)$.

Suppose first that $x \in \text{star}^V_{T-\{\{s,t\}\}}(s)$ and $y \notin \text{star}^V_{T-\{\{s,t\}\}}(s)$.
By Condition (4) of the definition of $d(T, \{s, t\})$ in 7.6,
$(d(T, \{s, t\}))(y, x) = 1$.
But by Part 2 and the Theorem of 2.16, either $s \in \text{star}^V_{Q-\{\{x,y\}\}}(x)$ and $t \notin \text{star}^V_{Q-\{\{x,y\}\}}(x)$, or $s \notin \text{star}^V_{Q-\{\{x,y\}\}}(x)$ and $t \in \text{star}^V_{Q-\{\{x,y\}\}}(x)$.
But by Part 3, $s \in \text{star}^V_{Q-\{\{x,y\}\}}(x)$, implying that $s \in \text{star}^V_{Q-\{\{x,y\}\}}(x)$ and $t \notin \text{star}^V_{Q-\{\{x,y\}\}}(x)$.
By Part 2 and Lemma 1, $s \notin \text{star}^V_{Q-\{\{x,y\}\}}(y)$ and $t \in \text{star}^V_{Q-\{\{x,y\}\}}(y)$.
By Part 2 and Condition (6) of the definition of $d(Q, \{x, y\})$ in 7.6,
$(d(Q, \{x, y\}))(s, t) = 1 = (d(T, \{s, t\}))(y, x)$.

On the other hand, suppose that $x \notin \text{star}^V_{T-\{\{s,t\}\}}(s)$ and $y \in \text{star}^V_{T-\{\{s,t\}\}}(s)$.
By Condition (3) of the definition of $d(T, \{s, t\})$ in 7.6,
$(d(T, \{s, t\}))(y, x) = -1$.
But by Part 2 and the Theorem of 2.16, either $s \in \text{star}^V_{Q-\{\{x,y\}\}}(y)$ and $t \notin \text{star}^V_{Q-\{\{x,y\}\}}(y)$, or $s \notin \text{star}^V_{Q-\{\{x,y\}\}}(y)$ and $t \in \text{star}^V_{Q-\{\{x,y\}\}}(y)$.
But by Part 4, $s \in \text{star}^V_{Q-\{\{x,y\}\}}(y)$, implying that $s \in \text{star}^V_{Q-\{\{x,y\}\}}(y)$ and $t \notin \text{star}^V_{Q-\{\{x,y\}\}}(y)$.
By Part 2 and Condition (5) of the definition of $d(Q, \{x, y\})$ in 7.6,
$(d(Q, \{x, y\}))(s, t) = -1 = (d(T, \{s, t\}))(y, x)$.

Case 4. $f(\{x, y\}) = (y, x)$ and $f(\{s, t\}) = (t, s)$.

By Part 1 and the Theorem of 2.16, either $x \in \text{star}^V_{T-\{\{s,t\}\}}(s)$ and $y \notin \text{star}^V_{T-\{\{s,t\}\}}(s)$, or $x \notin \text{star}^V_{T-\{\{s,t\}\}}(s)$ and $y \in \text{star}^V_{T-\{\{s,t\}\}}(s)$.

Suppose first that $x \in \text{star}^V_{T-\{\{s,t\}\}}(s)$ and $y \notin \text{star}^V_{T-\{\{s,t\}\}}(s)$.
By Lemma 1, $x \notin \text{star}^V_{T-\{\{s,t\}\}}(t)$ and $y \in \text{star}^V_{T-\{\{s,t\}\}}(t)$.
By Condition (5) of the definition of $d(T, \{s, t\})$ in 7.6,
$(d(T, \{s, t\}))(y, x) = -1$.
But by Part 2 and the Theorem of 2.16, either $s \in \text{star}^V_{Q-\{\{x,y\}\}}(x)$ and $t \notin \text{star}^V_{Q-\{\{x,y\}\}}(x)$, or $s \notin \text{star}^V_{Q-\{\{x,y\}\}}(x)$ and $t \in \text{star}^V_{Q-\{\{x,y\}\}}(x)$.
But by Part 3, $s \in \text{star}^V_{Q-\{\{x,y\}\}}(x)$, implying that $s \in \text{star}^V_{Q-\{\{x,y\}\}}(x)$ and $t \notin \text{star}^V_{Q-\{\{x,y\}\}}(x)$.
By Part 2 and Lemma 1, $s \notin \text{star}^V_{Q-\{\{x,y\}\}}(y)$ and $t \in \text{star}^V_{Q-\{\{x,y\}\}}(y)$.
By Part 2 and Condition (5) of the definition of $d(Q, \{x, y\})$ in 7.6,
$(d(Q, \{x, y\}))(t, s) = -1 = (d(T, \{s, t\}))(y, x)$.

On the other hand, suppose that $x \notin \text{star}^V_{T-\{\{s,t\}\}}(s)$ and $y \in \text{star}^V_{T-\{\{s,t\}\}}(s)$.
By Lemma 1, $x \in \text{star}^V_{T-\{\{s,t\}\}}(t)$ and $y \notin \text{star}^V_{T-\{\{s,t\}\}}(t)$.

By Condition (6) of the definition of $d(T, \{s, t\})$ in 7.6,
$$\big(d(T, \{s, t\})\big)(y, x) = 1.$$
But by Part 2 and the Theorem of 2.16, either $s \in \mathrm{star}^V_{Q-\{\{x,y\}\}}(y)$ and
$t \notin \mathrm{star}^V_{Q-\{\{x,y\}\}}(y)$, or $s \notin \mathrm{star}^V_{Q-\{\{x,y\}\}}(y)$ and $t \in \mathrm{star}^V_{Q-\{\{x,y\}\}}(y)$.
But by Part 4, $s \in \mathrm{star}^V_{Q-\{\{x,y\}\}}(y)$, implying that $s \in \mathrm{star}^V_{Q-\{\{x,y\}\}}(y)$
and $t \notin \mathrm{star}^V_{Q-\{\{x,y\}\}}(y)$.
By Part 2 and Condition (6) of the definition of $d(Q, \{x, y\})$ in 7.6,
$$\big(d(Q, \{x, y\})\big)(t, s) = 1 = \big(d(T, \{s, t\})\big)(y, x).$$

Lemma 3. *Let $T \in \mathscr{T}$. Let $(x, y) \in K$. Let $(s, t) \in K$.*
Then,
$$\big(B_T(\Phi^K_{(s,t)})\big)(x, y) \neq 0$$
if and only if
$$\{s, t\} \in T \text{ and } \{x, y\} = \{s, t\}$$
or
$$\{s, t\} \in T \text{ and } \{x, y\} \neq \{s, t\} \text{ and } \{x, y\} \in S - T \text{ and } \{s, t\} \in \lambda(T, \{x, y\}).$$

Proof: Suppose first that $\big(B_T(\Phi^K_{(s,t)})\big)(x, y) \neq 0$.

Now, if $\{s, t\} \in S - T$, then $B_T(\Phi^K_{(s,t)}) = \theta_1$ and $\big(B_T(\Phi^K_{(s,t)})\big)(x, y) = 0$,
which is false.
Thus, $\{s, t\} \in T$ and $B_T(\Phi^K_{(s,t)}) = -d(T, \{s, t\})$.

Now, if $\{x, y\} = \{s, t\}$ we are through.
Thus, assume that $\{x, y\} \neq \{s, t\}$.
By Condition (2) of the definition of $d(T, \{s, t\})$ in 7.6, we must have
$\{x, y\} \in \mu(T, \{s, t\})$.
But since $\{x, y\} \neq \{s, t\}$, Condition (2) of the definition of $\mu(T, \{s, t\})$
in 7.6 guarantees that $\{x, y\} \in S - T$.
By Condition (3) of the definition of $\mu(T, \{s, t\})$ in 7.6, $\{s, t\} \in \lambda(T, \{x, y\})$.

In the other direction, suppose first that $\{s, t\} \in T$ and $\{x, y\} = \{s, t\}$.
Then, $B_T(\Phi^K_{(s,t)}) = -d(T, \{s, t\})$, and by the Lemma of 8.6,
$$\big(B_T(\Phi^K_{(s,t)})\big)(x, y) = -\big(d(T, \{s, t\})\big)(s, t) = 1 \neq 0.$$

Finally, suppose that $\{s, t\} \in T$ and $\{x, y\} \neq \{s, t\}$ and $\{x, y\} \in S - T$
and $\{s, t\} \in \lambda(T\{x, y\})$.
Then, $B_T(\Phi^K_{(s,t)}) = -d(T, \{s, t\})$.
But by Condition (3) of the definition of $\mu(T, \{s, t\})$ in 7.6,
$\{x, y\} \in \mu(T, \{s, t\})$.
By Conditions (3)–(6) of the definition of $d(T, \{s, t\})$ in 7.6,
$\big|\big(B_T(\Phi^K_{(s,t)})\big)(x, y)\big| = \big|\big(d(T, \{s, t\})\big)(x, y)\big| = 1$, and thus,
$\big(B_T(\Phi^K_{(s,t)})\big)(x, y) \neq 0$.

Lemma 4. *Let $T \in \mathscr{T}$. Let $(x, y) \in K$. Let $(s, t) \in K$. Suppose that*
$\{x, y\} \neq \{s, t\}$. *Suppose that $\big(B_T(\Phi^K_{(s,t)})\big)(x, y) \neq 0$.*
Then,
$$\big(B_T(\Phi^K_{(x,y)})\big)(s, t) = 0.$$

12*

Proof: By Lemma 3, $\{x, y\} \in S - T$.

Thus, $B_T(\Phi_{(x,y)}^K) = \theta_1$, and $(B_T(\Phi_{(x,y)}^K))\,(s,t) = 0$.

Theorem 2. $Y \circ G = (Y \circ G)^t$.

Proof: We apply Theorem 1 of 4.7. Let $(x, y) \in K$ and let $(s, t) \in K$. It suffices to show that

$$\langle (Y \circ G)\,(\Phi_{(s,t)}^K),\, \Phi_{(x,y)}^K \rangle = \langle (Y \circ G)\,(\Phi_{(x,y)}^K),\, \Phi_{(s,t)}^K \rangle.$$

Now, if $\{x, y\} = \{s, t\}$, then $(x, y) = (s, t)$ and we are through. Thus, we may assume that $\{x, y\} \neq \{s, t\}$.

Now,

$$
\begin{aligned}
\langle (Y \circ G)\,(\Phi_{(s,t)}^K),\, \Phi_{(x,y)}^K \rangle &= \langle Y(G(\Phi_{(s,t)}^K)),\, \Phi_{(x,y)}^K \rangle \\
&= \langle Y((\sum_{T \in \mathcal{T}} U(T)\,B_T)\,(\Phi_{(s,t)}^K)),\, \Phi_{(x,y)}^K \rangle \\
&= \langle Y(\sum_{T \in \mathcal{T}} U(T)\,B_T(\Phi_{s,t)}^K)),\, \Phi_{x,y}^K \rangle \\
&= \langle \sum_{T \in \mathcal{T}} U(T)\,Y(B_T(\Phi_{(s,t)}^K)),\, \Phi_{(x,y)}^K \rangle \\
&= \sum_{T \in \mathcal{T}} U(T)\,\langle Y(B_T(\Phi_{(s,t)}^K)),\, \Phi_{(x,y)}^K \rangle.
\end{aligned}
$$

Similarly,

$$\langle (Y \circ G)\,(\Phi_{(x,y)}^K),\, \Phi_{(s,t)}^K \rangle = \sum_{T \in \mathcal{T}} U(T)\,\langle Y(B_T(\Phi_{(x,y)}^K)),\, \Phi_{(s,t)}^K \rangle.$$

Thus, to complete the proof we must show that

$$\sum_{T \in \mathcal{T}} U(T)\,\langle Y(B_T(\Phi_{(s,t)}^K)),\, \Phi_{(x,y)}^K \rangle = \sum_{T \in \mathcal{T}} U(T)\,\langle Y(B_T(\Phi_{(x,y)}^K)),\, \Phi_{(s,t)}^K \rangle.$$

To show these two sums equal we invoke the Assumption concerning summation introduced in 5.5.

In particular, it suffices to exhibit a univalent function h with dmn $h = \mathcal{T}$ and rng $h = \mathcal{T}$, such that for each $T \in \mathcal{T}$,

$$U(T)\,\langle Y(B_T(\Phi_{(s,t)}^K)),\, \Phi_{(x,y)}^K \rangle = U(h(T))\,\langle Y(B_{h(T)}(\Phi_{(x,y)}^K)),\, \Phi_{(s,t)}^K \rangle.$$

We proceed to specify this function h.

First note Lemma 4 which states that if $(B_T(\Phi_{(s,t)}^K))\,(x, y) \neq 0$, then $(B_T(\Phi_{(x,y)}^K))\,(s, t) = 0$, and also states that if $(B_T(\Phi_{x,y}^K))\,(s, t) \neq 0$, then $(B_T(\Phi_{(s,t)}^K))\,(x, y) = 0$.

Thus, to specify $h(T)$ for each $T \in \mathcal{T}$, it suffices first to specify $h(T)$ for $T \in \mathcal{T}$ such that $(B_T(\Phi_{(s,t)}^K))\,(x, y) \neq 0$, and then to specify $h(T)$ for $T \in \mathcal{T}$ such that $(B_T(\Phi_{(x,y)}^K))\,(s, t) \neq 0$, and finally to specify $h(T)$ for $T \in \mathcal{T}$ such that $(B_T(\Phi_{(s,t)}^K))\,(x, y) = 0$ and $(B_T(\Phi_{(x,y)}^K))\,(s, t) = 0$.

In particular, we let h be that function with domain $\mathscr{T}$ such that for each $T \in \mathscr{T}$,

$$h(T) = (T \cup \{\{x, y\}\}) - \{\{s, t\}\}, \text{ if } (B_T(\Phi^K_{(s,t)}))\,(x, y) \neq 0,$$
$$h(T) = (T \cup \{\{s, t\}\}) - \{\{x, y\}\}, \text{ if } (B_T(\Phi^K_{(x,y)}))\,(s, t) \neq 0,$$
$$h(T) = T, \text{ if } (B_T(\Phi^K_{(s,t)}))\,(x, y) = 0 = (B_T(\Phi^K_{(x,y)}))\,(s, t).$$

In the remainder of the proof which is divided into four parts we show that $\operatorname{rng} h = \mathscr{T}$, h is univalent, and for each $T \in \mathscr{T}$,

$$U(T)\,\langle Y(B_T(\Phi^K_{(s,t)})),\, \Phi^K_{(x,y)}\rangle = U(h(T))\,\langle Y(B_{h(T)}(\Phi^K_{(x,y)})),\, \Phi^K_{(s,t)}\rangle.$$

Part 1. $\operatorname{rng} h \subset \mathscr{T}$.

Proof of Part 1: Let $T \in \mathscr{T}$.

If $(B_T(\Phi^K_{(s,t)}))\,(x, y) \neq 0$, then $h(T) = (T \cup \{\{x, y\}\}) - \{\{s, t\}\}$, and by Lemma 3, $\{x, y\} \in S - T$ and $\{s, t\} \in \lambda(T, \{x, y\})$, and by Lemma 1, (2), of 8.3, $(T \cup \{\{x, y\}\}) - \{\{s, t\}\} \in \mathscr{T}$.

If $(B_T(\Phi^K_{(x,y)}))\,(s, t) \neq 0$, then $h(T) = (T \cup \{\{s, t\}\}) - \{\{x, y\}\}$, and by Lemma 3, $\{s, t\} \in S - T$ and $\{x, y\} \in \lambda(T, \{s, t\})$, and by Lemma 1, (2), of 8.3, $(T \cup \{\{s, t\}\}) - \{\{x, y\}\} \in \mathscr{T}$.

If $(B_T(\Phi^K_{(s,t)}))\,(x, y) = 0$ and $(B_T(\Phi^K_{(x,y)}))\,(s, t) = 0$, then $h(T) = T \in \mathscr{T}$.

Part 2. $\operatorname{rng} h = \mathscr{T}$.

Proof of Part 2: Let $Q \in \mathscr{T}$. We shall exhibit $T \in \mathscr{T}$ such that $h(T) = Q$.

Three cases arise.

Case 1. $(B_Q(\Phi^K_{(s,t)}))\,(x, y) \neq 0$.

In this case, let $T = (Q \cup \{\{x, y\}\}) - \{\{s, t\}\}$.

By Lemma 3, $\{x, y\} \in S - Q$ and $\{s, t\} \in \lambda(Q, \{x, y\})$.

Thus, by Lemma 1, (2), of 8.3, $(Q \cup \{\{x, y\}\}) - \{\{s, t\}\} \in \mathscr{T}$, and $T \in \mathscr{T}$.
We shall show that $h(T) = Q$.

By Lemma 1, (3), of 8.3, $\lambda(Q, \{x, y\}) = \lambda(T, \{s, t\})$.
But $\{x, y\} \in \lambda(Q, \{x, y\})$.
Thus, $\{x, y\} \in \lambda(T, \{s, t\})$.
But $\{x, y\} \in T$.
Also, $\{s, t\} \in S - T$.
Thus, by Lemma 3, $(B_T(\Phi^K_{(x,y)}))\,(s, t) \neq 0$.
By the definition of h,

$$\begin{aligned}
h(T) &= (T \cup \{\{s, t\}\}) - \{\{x, y\}\} \\
&= (((Q \cup \{\{x, y\}\}) - \{\{s, t\}\}) \cup \{\{s, t\}\}) - \{\{x, y\}\} \\
&= Q.
\end{aligned}$$

Case 2. $(B_Q(\Phi^K_{(x,y)}))\,(s,t) \ne 0$.

In this case, let $T = (Q \cup \{\{s,t\}\}) - \{\{x,y\}\}$.

By Lemma 3, $\{s,t\} \in S - Q$ and $\{x,y\} \in \lambda(Q, \{s,t\})$.

Thus, by Lemma 1, (2) of 8.3, $(Q \cup \{\{s,t\}\}) - \{\{x,y\}\} \in \mathcal{T}$, and $T \in \mathcal{T}$.

We shall show that $h(T) = Q$.

By Lemma 1, (3) of 8.3, $\lambda(Q, \{s,t\}) = \lambda(T, \{x,y\})$.

But $\{s,t\} \in \lambda(Q, \{s,t\})$.

Thus, $\{s,t\} \in \lambda(T, \{x,y\})$.

But $\{s,t\} \in T$.

Also, $\{x,y\} \in S - T$.

Thus, by Lemma 3, $(B_T(\Phi^K_{(s,t)}))\,(x,y) \ne 0$.

By the definition of h,

$$h(T) = (T \cup \{\{x,y\}\}) - \{\{s,t\}\}$$
$$= (((Q \cup \{\{s,t\}\}) - \{\{x,y\}\}) \cup \{\{x,y\}\}) - \{\{s,t\}\}$$
$$= Q.$$

Case 3. $(B_Q(\Phi^K_{(s,t)}))\,(x,y) = 0 = (B_Q(\Phi^K_{(x,y)}))\,(s,t)$.

In this case, let $T = Q$.

Then, by the definition of h, $h(T) = h(Q) = Q$.

Part 3. h is univalent.

Proof of Part 3: Let $T \in \mathcal{T}$ and let $Q \in \mathcal{T}$ such that $h(T) = h(Q)$.
We shall show that $T = Q$.

Because T and Q are arbitrary, it suffices to consider the following six cases.

Case 1. $(B_T(\Phi^K_{(s,t)}))\,(x,y) \ne 0$ and $(B_Q(\Phi^K_{(s,t)}))\,(x,y) \ne 0$.

In this case, since $h(T) = h(Q)$,

$$(T \cup \{\{x,y\}\}) - \{\{s,t\}\} = (Q \cup \{\{x,y\}\}) - \{\{s,t\}\}.$$

But by Lemma 3, $\{s,t\} \in T$ and $\{x,y\} \in S - T$.

Also, by Lemma 3, $\{s,t\} \in Q$ and $\{x,y\} \in S - Q$.

These facts imply that

$$T = (((T \cup \{\{x,y\}\}) - \{\{s,t\}\}) \cup \{\{s,t\}\}) - \{\{x,y\}\}$$
$$= (((Q \cup \{\{x,y\}\}) - \{\{s,t\}\}) \cup \{\{s,t\}\}) - \{\{x,y\}\}$$
$$= Q.$$

Case 2. $(B_T(\Phi^K_{(s,t)}))\,(x,y) \ne 0$ and $(B_Q(\Phi^K_{(x,y)}))\,(s,t) \ne 0$.

In this case, since $h(T) = h(Q)$,

$$(T \cup \{\{x,y\}\}) - \{\{s,t\}\} = (Q \cup \{\{s,t\}\}) - \{\{x,y\}\}.$$

But this is impossible, since $\{s, t\} \in (Q \cup \{\{s, t\}\}) - \{\{x, y\}\}$, but $\{s, t\} \notin (T \cup \{\{x, y\}\}) - \{\{s, t\}\}$.

Case 3. $(B_T(\Phi^K_{(s, t)}))\,(x, y) \neq 0$
and $(B_Q(\Phi^K_{(s, t)}))\,(x, y) = 0 = (B_Q(\Phi^K_{(x, y)}))\,(s, t).$

In this case, since $h(T) = h(Q)$,

$$(T \cup \{\{x, y\}\}) - \{\{s, t\}\} = Q.$$

By Lemma 3, $\{x, y\} \notin Q$ or $\{s, t\} \in Q$ or $\{x, y\} \notin \lambda(Q, \{s, t\})$.
Since $\{x, y\} \in (T \cup \{\{x, y\}\}) - \{\{s, t\}\} = Q$, it is not the case that $\{x, y\} \notin Q$.
If $\{s, t\} \in Q$, then $\{s, t\} \in (T \cup \{\{x, y\}\}) - \{\{s, t\}\}$, which is impossible.
Thus, we must have $\{x, y\} \notin \lambda(Q, \{s, t\})$.

But again, by Lemma 3, $\{x, y\} \in S - T$ and $\{s, t\} \in \lambda(T, \{x, y\})$.
By Lemma 1, (3), of 8.3, $\lambda(T, \{x, y\}) = \lambda(Q, \{s, t\})$.
But we have shown above that $\{x, y\} \notin \lambda(Q, \{s, t\})$.
Thus, $\{x, y\} \notin \lambda(T, \{x, y\})$, which is impossible.

Case 4. $(B_T(\Phi^K_{(x, y)}))\,(s, t) \neq 0$ and $(B_Q(\Phi^K_{(x, y)}))\,(s, t) \neq 0.$

This case is essentially the same as Case 1, except that the roles of $\{x, y\}$ and $\{s, t\}$ are interchanged.

Case 5. $(B_T(\Phi^K_{(x, y)}))\,(s, t) \neq 0$
and $(B_Q(\Phi^K_{(s, t)}))\,(x, y) = 0 = (B_Q(\Phi^K_{(x, y)}))\,(s, t).$

This case is essentially the same as Case 3, except that the roles of $\{x, y\}$ and $\{s, t\}$ are interchanged.

Case 6. $(B_T(\Phi^K_{(s, t)}))\,(x, y) = 0 = (B_T(\Phi^K_{(x, y)}))\,(s, t)$
and $(B_Q(\Phi^K_{(s, t)}))\,(x, y) = 0 = (B_Q(\Phi^K_{(x, y)}))\,(s, t).$

In this case, $T = h(T) = h(Q) = Q.$

Part 4. $T \in \mathcal{F}$
implies
$$U(T)\,\langle Y(B_T(\Phi^K_{(s, t)})),\, \Phi^K_{(x, y)}\rangle = U(h(T))\,\langle Y(B_{h(T)}(\Phi^K_{(x, y)})),\, \Phi^K_{(s, t)}\rangle.$$

Proof of Part 4: Let $T \in \mathcal{F}$.
Note that

$$\langle Y(B_T(\Phi^K_{(s, t)})),\, \Phi^K_{(x, y)}\rangle = \langle Y(\sum_{(u, v) \in K} (B_T(\Phi^K_{(s, t)}))\,(u, v)\,\Phi^K_{(u, v)}),\, \Phi^K_{(x, y)}\rangle$$

$$= \langle \sum_{(u, v) \in K} (B_T(\Phi^K_{(s, t)}))\,(u, v)\,Y(\Phi^K_{(u, v)}),\, \Phi^K_{(x, y)}\rangle$$

$$= \langle \sum_{(u, v) \in K} (B_T(\Phi^K_{(s, t)}))\,(u, v)\,(r(\{u, v\}))^{-1}\,\Phi^K_{(u, v)},\, \Phi^K_{(x, y)}\rangle$$

$$= \sum_{(u, v) \in K} (B_T(\Phi^K_{(s, t)}))\,(u, v)\,(r(\{u, v\}))^{-1}\,\langle \Phi^K_{(u, v)},\, \Phi^K_{(x, y)}\rangle$$

$$= (B_T(\Phi^K_{(s, t)}))\,(x, y)\,(r(\{x, y\}))^{-1}.$$

Similarly,

$$\langle Y(B_{h(T)}(\Phi^K_{(x,y)})), \Phi^K_{(s,t)} \rangle = (B_{h(T)}(\Phi^K_{(x,y)}))(s,t)(r(\{s,t\}))^{-1}.$$

Thus, it suffices to show that

$$U(T)(r(\{x,y\}))^{-1}(B_T(\Phi^K_{(s,t)}))(x,y)$$
$$= U(h(T))(r(\{s,t\}))^{-1}(B_{h(T)}(\Phi^K_{(x,y)}))(s,t).$$

Three cases arise.

Case 1. $(B_T(\Phi^K_{(s,t)}))(x,y) \neq 0.$

In this case, by Lemma 3, $\{s,t\} \in T$ and $B_T(\Phi^K_{(s,t)}) = -d(T,\{s,t\}).$
Thus,

$$U(T)(r(\{x,y\}))^{-1}(B_T(\Phi^K_{(s,t)}))(x,y)$$
$$= -(r(\{x,y\}))^{-1}\prod_{\{u,v\}\in T}(r(\{u,v\}))^{-1}(d(T,\{s,t\}))(x,y)$$
$$= -\prod_{\{u,v\}\in T \cup \{\{x,y\}\}}(r(\{u,v\}))^{-1}(d(T,\{s,t\}))(x,y).$$

On the other hand,

$$h(T) = (T \cup \{\{x,y\}\}) - \{\{s,t\}\}.$$

Thus, $\{x,y\} \in h(T)$, and $B_{h(T)}(\Phi^K_{(x,y)}) = -d(h(T),\{x,y\}).$ Hence,

$$U(h(T))(r(\{s,t\}))^{-1}(B_{h(T)}(\Phi^K_{(x,y)}))(s,t)$$
$$= -(r(\{s,t\}))^{-1}\prod_{\{u,v\}\in h(T)}(r(\{u,v\}))^{-1}(d(h(T),\{x,y\}))(s,t)$$
$$= -\prod_{\{u,v\}\in h(T)\cup\{\{s,t\}\}}(r(\{u,v\}))^{-1}(d(h(T),\{x,y\}))(s,t).$$

But by Lemma 3, $\{x,y\} \in S - T$ and $\{s,t\} \in \lambda(T,\{x,y\}).$
By Lemma 2,

$$(d(T,\{s,t\}))(x,y) = (d(h(T),\{x,y\}))(s,t).$$

Thus, to complete this case, it suffices to show that

$$T \cup \{\{x,y\}\} = h(T) \cup \{\{s,t\}\},$$

which is true because $\{s,t\} \in T.$

Case 2. $(B_T(\Phi^K_{(x,y)}))(s,t) \neq 0.$

In this case, by Lemma 4, $(B_T(\Phi^K_{(s,t)}))(x,y) = 0$, and

$$U(T)(r(\{x,y\}))^{-1}(B_T(\Phi^K_{(s,t)}))(x,y) = 0.$$

Thus, it suffices to show that $(B_{h(T)}(\Phi^K_{(x,y)}))(s,t) = 0.$
But,

$$\{x,y\} \notin (T \cup \{\{s,t\}\}) - \{\{x,y\}\} = h(T)$$

Hence, $B_{h(T)}(\Phi^K_{(x,y)}) = \theta_1$ and $(B_{h(T)}(\Phi^K_{(x,y)}))\,(s,t) = 0$.

Case 3. $(B_T(\Phi^K_{(s,t)}))\,(x,y) = 0 = (B_T(\Phi^K_{(x,y)}))\,(s,t)$.

In this case,

$$U(T)\,(r(\{x,y\}))^{-1}\,(B_T(\Phi^K_{(s,t)}))\,(x,y) = 0,$$

and it suffices to show that $(B_{h(T)}(\Phi^K_{(x,y)}))\,(s,t) = 0$, which is true since $h(T) = T$.

The proof is complete.

8.8 The Sum of Tree Branch Products

The sum of all tree branch products, $\sum\limits_{T \in \mathscr{T}} U(T)$, will occur sufficiently often for us to endow it with a special symbol. In particular, we shall let J be equal to the sum of the tree branch products. Formally,

$$J = \sum_{T \in \mathscr{T}} U(T).$$

Note from the definition of U in 8.7 that J is a positive real number.

The tree branch map G, applied to any coboundary, merely yields the number J multiplied by this coboundary. We exhibit this fact as the following Theorem.

Theorem. *Let $u \in \mathrm{rng}\ \delta$.*

Then,

$G(u) = J\,u$.

Proof: By Theorem 2 of 8.6,

$$\begin{aligned}
G(u) &= \Big(\sum_{T \in \mathscr{T}} U(T)\,B_T\Big)(u) \\
&= \sum_{T \in \mathscr{T}} U(T)\,B_T(u) \\
&= \sum_{T \in \mathscr{T}} U(T)\,u \\
&= \Big(\sum_{T \in \mathscr{T}} U(T)\Big)\,u \\
&= J\,u.
\end{aligned}$$

As an immediate Corollary to the preceding Theorem we exhibit the fact that the restriction of the composition function $Y \circ G$ to the coboundary space, range δ, is equal to the restriction of the diagonal linear map $J\,Y$ to the coboundary space, range δ.

Corollary. $(Y \circ G)|\mathrm{rng}\ \delta = (J\,Y)|\mathrm{rng}\ \delta$.

Proof: Let $u \in \mathrm{rng}\ \delta$. It suffices to show that $(Y \circ G)\,(u) = J\,Y(u)$. But by the preceding Theorem,

$$(Y \circ G)\,(u) = Y(G(u)) = Y(J\,u) = J\,Y(u).$$

8.9 The Voltage Chain with Current Sources

We are now in a position to specify the voltage chain $v \in \mathscr{L}(K)$ when our resistive network is energized exclusively with current sources. Recall, however, that one representation of the voltage chain has already been exhibited in Theorem 1 of 7.4, establishing the existence of the voltage chain and current chain satisfying Ohm's Law, Kirchhoff's Voltage Law, and Kirchhoff's Current Law. In this Chapter we seek another representation for the voltage chain $v \in \mathscr{L}(K)$ which will allow us to calculate easily the branch voltage $v(x, y)$ for each directed branch (x, y) of K.

To make matters precise, let the current source chain be $q \in \mathscr{L}(K)$. By Theorem 1 of 7.4, we are assured of the existence of a voltage chain $v \in \mathscr{L}(K)$ such that v is a coboundary and $Y(v) - q$ is a cycle. Thus,

$$v \in \mathrm{rng}\ \delta,$$
$$Y(v) - q \in \mathrm{kernel}\ \varDelta.$$

Before exhibiting the formula for the branch voltage $v(x, y)$ for each directed branch (x, y) of K, we need a preliminary Theorem 1 providing us with the value of the diagonal linear map $J\,Y$ evaluated at v.

Theorem 1. *Let $q \in \mathscr{L}(K)$. Let $v \in \mathrm{rng}\ \delta$. Suppose that* $Y(v) - q \in \mathrm{kernel}\ \varDelta$.
Then,
$$J\,Y(v) = \sum_{(s,t) \in K} q(s,t)\, G^t(\varPhi^K_{(s,t)}).$$

Proof: By Theorem 1 of 8.7, $G^t(Y(v) - q) = \theta_1$. Thus, $G^t(Y(v)) - G^t(q) = G^t(Y(v) - q) = \theta_1$, implying that $G^t(Y(v)) = G^t(q)$.
But by Theorem 8 of 4.7, $Y^t = Y$.
Thus, by Theorem 7 of 4.7,

$$G^t(q) = G^t(Y(v)) = (G^t \circ Y)(v) = (G^t \circ Y^t)(v) = (Y \circ G)^t(v).$$

Hence, by Theorem 2 of 8.7 and the Corollary of 8.8,

$$G^t(q) = (Y \circ G)^t(v) = (Y \circ G)(v) = J\,Y(v).$$

But also,

$$G^t(q) = G^t\Big(\sum_{(s,t) \in K} q(s,t)\, \varPhi^K_{(s,t)}\Big) = \sum_{(s,t) \in K} q(s,t)\, G^t(\varPhi^K_{(s,t)}).$$

Thus, the proof is complete.

As the following Theorem 2 we can now specify a usable formula for the branch voltage $v(x, y)$ for each directed branch (x, y) of K. The formula exhibited in Theorem 2 is frequently called Kirchhoff's Fourth Law.

Theorem 2. *Let* $q \in \mathscr{L}(K)$. *Let* $v \in \mathrm{rng}\ \delta$. *Suppose that*
$Y(v) - q \in \mathrm{kernel}\ \Delta$. *Let* $(x, y) \in K$.
Then,

$$v(x, y) = J^{-1} r(\{x, y\}) \sum_{(s,t) \in K} q(s, t) \sum_{T \in \mathscr{T}} U(T)\, (B_T(\Phi^K_{(x,y)}))\, (s, t).$$

Proof: Note that

$$J\,Y(v) = J\,Y\Big(\sum_{(s,t) \in K} v(s, t)\, \Phi^K_{(s,t)} \Big)$$

$$= J \sum_{(s,t) \in K} v(s, t)\, Y(\Phi^K_{(s,t)})$$

$$= J \sum_{(s,t) \in K} v(s, t)\, (r(\{s, t\}))^{-1}\, \Phi^K_{(s,t)}.$$

Thus,

$$\langle J\,Y(v), \Phi^K_{(x,y)} \rangle = \langle J \sum_{(s,t) \in K} v(s, t)\, (r(\{s, t\}))^{-1}\, \Phi^K_{(s,t)}, \Phi^K_{(x,y)} \rangle$$

$$= J \sum_{(s,t) \in K} v(s, t)\, (r(\{s, t\}))^{-1}\, \langle \Phi^K_{(s,t)}, \Phi^K_{(x,y)} \rangle$$

$$= J\,(r(\{x, y\}))^{-1}\, v(x, y).$$

Thus,

$$v(x, y) = J^{-1}\, r(\{x, y\})\, \langle J\,Y(v), \Phi^K_{(x,y)} \rangle,$$

and to evaluate $\langle J\,Y(v), \Phi^K_{(x,y)} \rangle$ we use Theorem 1.

Noting Theorem 2 of 4.7,

$$\langle J\,Y(v), \Phi^K_{(x,y)} \rangle = \langle \sum_{(s,t) \in K} q(s, t)\, G^t(\Phi^K_{(s,t)}), \Phi^K_{(x,y)} \rangle$$

$$= \sum_{(s,t) \in K} q(s, t)\, \langle G^t(\Phi^K_{(s,t)}), \Phi^K_{(x,y)} \rangle$$

$$= \sum_{(s,t) \in K} q(s, t)\, \langle G(\Phi^K_{(x,y)}), \Phi^K_{(s,t)} \rangle$$

$$= \sum_{(s,t) \in K} q(s, t)\, (G(\Phi^K_{(x,y)}))\, (s, t)$$

$$= \sum_{(s,t) \in K} q(s, t)\, \big(\big(\sum_{T \in \mathscr{T}} U(T)\, B_T \big)\, (\Phi^K_{(x,y)})\big)\, (s, t)$$

$$= \sum_{(s,t) \in K} q(s, t)\, \big(\sum_{T \in \mathscr{T}} U(T)\, B_T(\Phi^K_{(x,y)}) \big)\, (s, t)$$

$$= \sum_{(s,t) \in K} q(s, t) \sum_{T \in \mathscr{T}} U(T)\, (B_T(\Phi^K_{(x,y)}))\, (s, t).$$

Hence, since

$$v(x, y) = J^{-1}\, r(\{x, y\})\, \langle J\,Y(v), \Phi^K_{(x,y)} \rangle,$$

we obtain the desired result,

$$v(x, y) = J^{-1}\, r(\{x, y\}) \sum_{(s,t) \in K} q(s, t) \sum_{T \in \mathscr{T}} U(T)\, (B_T(\Phi^K_{(x,y)}))\, (s, t).$$

The result of the preceding Theorem 2 does not appear, at first glance, to offer an easily calculated branch voltage $v(x, y)$ for each directed branch (x, y) of K. However, to analyze this formula we consider a special case of the current source chain $q \in \mathscr{L}(K)$ when q is such that it assigns a zero source current to all directed branches of K except at most one. Such a source chain can be written as an element $m \, \Phi^K_{(s, t)}$ of $\mathscr{L}(K)$ where m is a real number and (s, t) is a directed branch of K.

As the following Corollary we reduce the formula exhibited in Theorem 2 to the special case of this source chain $m \, \Phi^K_{(s, t)}$. Since the Corollary follows trivially from the preceding Theorem 2, we omit the proof.

Corollary. *Let* $m \in R$. *Let* $(s, t) \in K$. *Let* $v \in \mathrm{rng}\,\delta$. *Suppose that* $Y(v) - m \, \Phi^K_{(s, t)} \in$ kernel Δ. *Let* $(x, y) \in K$.
Then,

$$v(x, y) = J^{-1}\, r(\{x, y\})\, m \sum_{T \in \mathscr{T}} U(T)\, (B_T(\Phi^K_{(x, y)}))\, (s, t).$$

Now, returning again to the general source chain $q \in \mathscr{L}(K)$, we can write,

$$q = \sum_{(s, t) \in K} q(s, t)\, \Phi^K_{(s, t)},$$

and thus, q is a sum of the special type of source chains considered in the Corollary, each of which assigns a non-zero source current to at most one directed branch of K. Observe that the branch voltage $v(x, y)$ resulting from this general source chain $q \in \mathscr{L}(K)$ is the sum of the individual branch voltages, each resulting from consideration of the special type of source chains used in the Corollary. Thus, in loose language, we can state that the effect of sources upon the branch voltage $v(x, y)$ is additive, and to investigate the formula for the branch voltage, we can concentrate upon the representation exhibited in the Corollary, in the case of the special source chain which assigns a non-zero source current to at most one directed branch of $\mathscr{L}(K)$.

Now, those trees $T \in \mathscr{T}$ which contribute a non-zero term to the formula for $v(x, y)$ exhibited in the Corollary are precisely those trees $T \in \mathscr{T}$ such that

$$(B_T(\Phi^K_{(x, y)}))\, (s, t) \neq 0.$$

Thus, Lemma 3 of 8.7 shows that we need only consider those trees $T \in \mathscr{T}$ such that either $\{x, y\} \in T$ and $\{x, y\} = \{s, t\}$, or $\{x, y\} \in T$ and $\{x, y\} \neq \{s, t\}$ and $\{s, t\} \in S - T$ and $\{x, y\} \in \lambda(T, \{s, t\})$. This simple observation reduces tremendously the number of trees to be considered, for if $\{x, y\} = \{s, t\}$ we consider only those trees $T \in \mathscr{T}$, such that $\{x, y\}$ is a branch of T, while if $\{x, y\} \neq \{s, t\}$, we only consider those trees $T \in \mathscr{T}$ such that $\{x, y\}$ is a branch of T, $\{s, t\}$ is not a branch of T, and

$\{x, y\}$ is a branch of the unique loop formed by adjoining $\{s, t\}$ to the tree. This reduction in the number of trees to be considered makes the formula exhibited in the Corollary easily computable.

8.10 Invariance Under Change of Incidence

Suppose that our resistive network is presented to an electrical engineer to be used in an actual circuit. It is stylish to describe the network as a closed black box, with certain external connections which are to be connected to the larger, more complicated circuit. Observe that the internal structure of the black box merely consists of a collection of resistors wired together in some known way. When this black box is presented to the electrical engineer, no directions have been assigned to the branches inside of the box.

As soon as this black box, with no directions assigned to its internal branches, is inserted into the larger, more complicated circuit, branch voltages and branch currents develop in all branches of the black box, satisfying Ohm's Law, Kirchhoff's Voltage Law, and Kirchhoff's Current Law. Note, however, that these actual branch voltages and branch currents are developed, in spite of the fact that no direction has been assigned to any of the branches inside the black box.

On the other hand, in our abstract setting we have developed elaborate formulas describing the branch voltages and branch currents induced in the branches of the network, but in order to develop our formulas we found it necessary to assign a direction to each branch of the network by means of the incidence function f. It appears, therefore, that our formulas for the branch voltages and branch currents induced in the branches of the network may depend upon the incidence function f used in their development. However, such a dependence upon the incidence function f would contradict the known electrical engineering situation, since no directions are assigned to the branches inside of the black box when it is inserted into the larger, more complicated circuit. Thus, to show that our abstract results do conform to the common electrical engineering situation, we must show that our formulas for the branch voltages and branch currents are essentially independent of the incidence function f, and we must describe this essential independence in a way which must be made precise.

To pin matters down, suppose that g is another incidence function of our set of branches S and let $\{x, y\}$ be a branch of S. We must compare the branch voltage and branch current developed from the incidence function f and evaluated at the directed branch $f(\{x, y\})$ with the branch voltage and branch current developed from the incidence function g

evaluated at the directed branch $g(\{x, y\})$. Such a comparison demands that we develop a corresponding algebraic and topological structure for the incidence function g in order to evaluate the branch voltage and branch current developed from the incidence function g at the directed branch $g(\{x, y\})$.

To compare the branch voltage and branch current developed from the incidence function f and evaluated at the directed branch $f(\{x, y\})$ with the branch voltage and branch current developed from the incidence function g and evaluated at the directed branch $g(\{x, y\})$, we must first see whether f and g assign the same or different directions to the branch $\{x, y\}$. If f and g both assign the same direction to the branch $\{x, y\}$ we expect the branch voltage and branch current developed from the incidence function f and evaluated at the directed branch $f(\{x, y\})$ to be equal to the corresponding branch voltage and branch current developed from the incidence function g and evaluated at the directed branch $g(\{x, y\})$. On the other hand, if f and g assign different directions to the branch $\{x, y\}$ we expect the branch voltage and branch current developed from the incidence function f and evaluated at the directed branch $f(\{x, y\})$ to be the negative of the corresponding branch voltage and branch current developed from the incidence function g and evaluated at the directed branch $g(\{x, y\})$.

We describe the preceding relations between the branch voltages and branch currents developed from the two incidence functions f and g by saying that the branch voltages and branch currents developed from the two incidence functions are essentially the same. Thus, if we can show that our expectations described above are indeed true, then we can assert that our abstract formulas developed for the branch voltages and branch currents are essentially independent of the incidence function used in their construction.

Consider the new incidence function g, which may differ from f by assigning to each of several branches of S a direction different from the direction assigned to the branch by f. Starting with the incidence function f, we can manufacture the incidence function g by considering a finite sequence of incidence functions, each of which differs from the preceding incidence function of the sequence only in the direction assigned to exactly one branch of S. If we can show that the corresponding branch voltages and branch currents developed from two consecutive incidence functions of the sequence are essentially the same, it will follow that the corresponding branch voltages and branch currents developed from the incidence functions f and g are essentially the same.

Hence, we can simplify the problem by considering as our second incidence function, a new incidence function f' with domain equal to S, such that f' agrees with f at all branches of S except one. We must show that the corresponding branch voltages and branch currents developed from the incidence functions f and f' are essentially the same.

To make matters precise, let $\{u, z\}$ be a branch of S and suppose that $f(\{u, z\}) = (u, z)$. We let f' be that function with domain equal to S such that

$$f'(\{s, t\}) = f(\{s, t\}) \text{ if } \{s, t\} \in S \text{ and } \{s, t\} \neq \{u, z\}.$$
$$f'(\{u, z\}) = (z, u).$$

The algebraic and topological structures must now be specified for the incidence function f'. In particular, we let

$$K' = \mathrm{rng}\, f',$$

and we let Δ' and δ' be the boundary and coboundary operators associated with f' corresponding to the boundary and coboundary operators Δ and δ associated with f.

We let Z' be the impedance function associated with f' corresponding to the impedance function Z associated with f. We let Y' be the linear map of $\mathscr{L}(K')$ into $\mathscr{L}(K')$ associated with f' corresponding to the linear map Y of $\mathscr{L}(K)$ into $\mathscr{L}(K)$ associated with f.

For each $T \in \mathscr{T}$ and each branch $\{s, t\} \in S - T$ we let $c'(T, \{s, t\})$ be the cycle associated with f' corresponding to the cycle $c(T, \{s, t\})$ associated with f. For each $T \in \mathscr{T}$ and each branch $\{s, t\} \in T$ we let $d'(T, \{s, t\})$ be the coboundary associated with f' corresponding to the coboundary $d(T, \{s, t\})$ associated with f.

For each $T \in \mathscr{T}$ we let A'_T be the cycle map of $\mathscr{L}(K')$ into $\mathscr{L}(K')$ associated with f' corresponding to the cycle map A_T associated with f. For each $T \in \mathscr{T}$ we let B'_T be the coboundary map of $\mathscr{L}(K')$ into $\mathscr{L}(K')$ associated with f' corresponding to the coboundary map B_T associated with f.

We must consider separately the cases of excitation exclusively by voltage sources or excitation exclusively by current sources. We first consider excitation exclusively by voltage sources and let $e \in \mathscr{L}(K)$ be the voltage source chain associated with the incidence function f energizing the network. To compare the branch voltages and branch currents developed from the incidence function f with the branch voltages and branch currents developed from the incidence function f', we must now assume a voltage source chain $e' \in \mathscr{L}(K')$ associated with the incidence function f' which corresponds to the voltage source chain $e \in \mathscr{L}(K)$ associated with the incidence function f. Furthermore, the voltage source chain $e' \in \mathscr{L}(K')$ associated with the incidence function f' must deliver to the network exactly the same external excitation as the voltage chain $e \in \mathscr{L}(K)$ associated with the incidence function f. Thus, we require that e' is a function with domain equal to K' such that

$$e'(s, t) = e(s, t) \text{ if } (s, t) \in K' \text{ and } (s, t) \neq (z, u),$$

$$e'(z, u) = -e(u, z).$$

Next, we let $i' \in \mathscr{L}(K')$ be the branch current chain associated with the incidence function f' corresponding to the branch current chain $i \in \mathscr{L}(K)$ associated with the incidence function f. Since Ohm's Law, Kirchhoff's Voltage Law, and Kirchhoff's Current Law must be satisfied, we can assert that

$$i' \in \text{kernel } \Delta',$$

$$Z'(i') - e' \in \text{rng } \delta'.$$

We fix some branch $\{x, y\} \in S$ and we want to compare $i(f(\{x, y\}))$ with $i'(f'(\{x, y\}))$. In particular, we expect to show that if $\{x, y\} \neq \{u, z\}$, then $i(f(\{x, y\})) = i'(f'(\{x, y\}))$, while if $\{x, y\} = \{u, z\}$, then $i(f(\{x, y\})) = -i'(f'(\{x, y\}))$.

From Theorem 2 of 8.4, we know that

$$i(f(\{x, y\})) = D^{-1}(r(\{x, y\}))^{-1} \sum_{(s, t) \in K} e(s, t) \sum_{T \in \mathscr{T}} W(T) \left(A_T(\Phi^K_{f(\{x, y\})})\right)(s, t)$$

$$= D^{-1}(r(\{x, y\}))^{-1} \sum_{T \in \mathscr{T}} W(T) \sum_{(s, t) \in K} e(s, t) \left(A_T(\Phi^K_{f(\{x, y\})})\right)(s, t).$$

Similarly,

$$i'(f'(\{x, y\})) = D^{-1}(r(\{x, y\}))^{-1} \sum_{T \in \mathscr{T}} W(T) \sum_{(s, t) \in K'} e'(s, t) \left(A'_T(\Phi^{K'}_{f'(\{x, y\})})\right)(s, t).$$

Thus, to compare $i(f(\{x, y\}))$ with $i'(f'(\{x, y\}))$, we select some fixed $T \in \mathscr{T}$, and compare

$$\sum_{(s, t) \in K} e(s, t) \left(A_T(\Phi^K_{f(\{x, y\})})\right)(s, t)$$

with

$$\sum_{(s, t) \in K'} e'(s, t) \left(A'_T(\Phi^{K'}_{f'(\{x, y\})})\right)(s, t).$$

Now,

$$\sum_{(s, t) \in K} e(s, t) \left(A_T(\Phi^K_{f(\{x, y\})})\right)(s, t)$$

$$= \sum_{\{s, t\} \in S} e(f(\{s, t\})) \left(A_T(\Phi^K_{f(\{x, y\})})\right)(f(\{s, t\}))$$

and

$$\sum_{(s, t) \in K'} e'(s, t) \left(A'_T(\Phi^{K'}_{f'(\{x, y\})})\right)(s, t)$$

$$= \sum_{\{s, t\} \in S} e'(f'(\{s, t\})) \left(A'_T(\Phi^{K'}_{f'(\{x, y\})})\right)(f'(\{s, t\})).$$

Thus, we must compare

$$\sum_{\{s,\,t\}\in S} e(f(\{s,t\}))\,(A_T(\Phi^K_{f(\{x,y\})}))\,(f(\{s,t\}))$$

with

$$\sum_{\{s,\,t\}\in S} e'(f'(\{s,t\}))\,(A'_T(\Phi^{K'}_{f'(\{x,y\})}))\,(f'(\{s,t\}))\,.$$

To do this, select some fixed $\{s,t\}\in S$.

We shall show that if $\{x,y\}\neq\{u,z\}$,

$$e(f(\{s,t\}))\,(A_T(\Phi^K_{f(\{x,y\})}))\,(f(\{s,t\})$$
$$= e'(f'(\{s,t\}))\,(A'_T(\Phi^{K'}_{f'(\{x,y\})}))\,(f'(\{s,t\}))\,,$$

while if $\{x,y\}=\{u,z\}$,

$$e(f(\{s,t\}))\,(A_T(\Phi^K_{f(\{x,y\})}))\,(f(\{s,t\}))$$
$$= -\,e'(f'(\{s,t\}))\,(A'_T(\Phi^{K'}_{f'(\{x,y\})}))\,(f'(\{s,t\}))\,.$$

Now, by Lemma 4 of 8.3, if $\{x,y\}\in T$ or $\{s,t\}\notin\lambda(T,\{x,y\})$, then

$$(A_T(\Phi^K_{f(\{x,y\})}))\,(f(\{s,t\})) = 0 = (A'_T(\Phi^{K'}_{f'(\{x,y\})}))\,(f'(\{s,t\}))\,,$$

and the desired result is achieved.

Thus, we may assume that $\{x,y\}\in S-T$ and $\{s,t\}\in\lambda(T,\{x,y\})$, implying that $A_T(\Phi^K_{f(\{x,y\})}) = c(T,\{x,y\})$ and $A'_T(\Phi^{K'}_{f'(\{x,y\})}) = c'(T,\{x,y\})$. Thus, we must show that if $\{x,y\}\neq\{u,z\}$,

$$e(f(\{s,t\}))\,(c(T,\{x,y\}))\,(f(\{s,t\}))$$
$$= e'(f'(\{s,t\}))\,(c'(T,\{x,y\}))\,(f'(\{s,t\}))\,,$$

while if $\{x,y\}=\{u,z\}$,

$$e(f(\{s,t\}))\,(c(T,\{x,y\}))\,(f(\{s,t\}))$$
$$= -\,e'(f'(\{s,t\}))\,(c'(T,\{x,y\}))\,(f'(\{s,t\}))\,.$$

Two cases arise.

Case 1. $\{x,y\}\neq\{u,z\}$.

In this case, we must consider separately whether $\{s,t\}\neq\{u,z\}$ or $\{s,t\}=\{u,z\}$.

Thus, we suppose first that $\{s,t\}\neq\{u,z\}$, implying that $f'(\{s,t\})=f(\{s,t\})$ and $e'(f'(\{s,t\}))=e(f(\{s,t\}))$.

Now, either $\{s,t\}=\{x,y\}$ or $\{s,t\}\neq\{x,y\}$.

If $\{s,t\}=\{x,y\}$, then

$$(c(T,\{x,y\}))\,(f(\{s,t\})) = 1 = (c'(T,\{x,y\}))\,(f(\{s,t\}))\,,$$

leading to the desired result.

If $\{s, t\} \neq \{x, y\}$, then $\{s, t\} \in \lambda(T, \{x, y\}) - \{\{x, y\}\}$, and Lemma 2 of 8.3 shows that

$$(c(T(\{x, y\}))) (f(\{s, t\})) = (c'(T, \{x, y\})) (f(\{s, t\}))),$$

leading to the desired result.

Suppose now that $\{s, t\} = \{u, z\}$, implying that $f(\{s, t\}) = (u, z)$ and $f'(\{s, t\}) = (z, u)$, and thus, $e'(f'(\{s, t\})) = - e(f(\{s, t\}))$. But $\{s, t\} = \{u, z\}$ also implies that $\{s, t\} \neq \{x, y\}$ and $\{s, t\} \in \lambda(T, \{x, y\}) - \{\{x, y\}\}$, so that by Lemma 2 of 8.3,

$$(c(T, \{x, y\})) (f(\{s, t\})) = - (c'(T, \{x, y\})) (f'(\{s, t\})).$$

Combining this fact with the assertion above that $e'(f'(\{s, t\})) = - e(f(\{s, t\}))$ gives the desired result.

Case 2. $\{x, y\} = \{u, z\}$.

Suppose first that $\{s, t\} \neq \{u, z\}$, implying that $f(\{s, t\}) = f'(\{s, t\})$ and $e(f(\{s, t\})) = e'(f'(\{s, t\}))$. Also, $\{s, t\} \neq \{x, y\}$ and $\{s, t\} \in \lambda(T, \{x, y\}) - \{\{x, y\}\}$, so that by Lemma 2 of 8.3,

$$(c(T, \{x, y\})) (f(\{s, t\})) = - (c'(T, \{x, y\})) (f(\{s, t\})),$$

leading to the desired result.

Finally, if $\{s, t\} = \{u, z\}$, then $f(\{s, t\}) = (u, z)$ and $f'(\{s, t\}) = (z, u)$, and thus $e'(f'(\{s, t\})) = - e(f(\{s, t\}))$. But with $\{s, t\} = \{u, z\}$, $\{s, t\} = \{x, y\}$, and

$$(c(T, \{x, y\})) (f(\{s, t\})) = 1 = (c'(T, \{x, y\})) (f'(\{s, t\})),$$

which, combined with the assertion above that $e'(f'(\{s, t\})) = - e(f(\{s, t\}))$, gives the desired result.

Next, we let $v' \in \mathscr{L}(K')$ be the branch voltage chain associated with the incidence function f' corresponding to the branch voltage chain $v \in \mathscr{L}(K)$ associated with the incidence function f. Observe that both v and v' satisfy Ohm's Law, and thus, $v = Z(i)$ and $v' = Z'(i')$. With our fixed branch $\{x, y\} \in S$ we expect to show that if $\{x, y\} \neq \{u, z\}$, then $v(f(\{x, y\})) = v'(f'(\{x, y\}))$, while if $\{x, y\} = \{u, z\}$, then $v(f(\{x, y\})) = - v'(f'(\{x, y\}))$.

But from the discussion of Ohm's Law in 6.4, we know that the fact that $v = Z(i)$ implies that

$$v(f(\{x, y\})) = r(\{x, y\}) \, i(f(x, y\})),$$

while the fact that $v' = Z'(i')$ implies that

$$v'(f'(\{x, y\})) = r(\{x, y\}) \, i'(f'(\{x, y\})).$$

Thus, from our previous facts relating $i(f(\{x, y\}))$ and $i'(f'(\{x, y\}))$, we obtain the desired results relating $v(f(\{x, y\}))$ and $v'(f'(\{x, y\}))$.

We now consider excitation exclusively by current sources and let $q \in \mathscr{L}(K)$ be the current source chain associated with the incidence function f energizing the network. To compare the branch voltages and branch currents developed from the incidence function f with the branch voltages and branch currents developed from the incidence function f', we must now assume a current source chain $q' \in \mathscr{L}(K')$ associated with the incidence function f' which corresponds to the current source chain $q \in \mathscr{L}(K)$ associated with the incidence function f. Furthermore, the current source chain $q' \in \mathscr{L}(K')$ associated with the incidence function f' must deliver to the network exactly the same external excitation as the current source chain $q \in \mathscr{L}(K)$ associated with the incidence function f. Thus, we require that q' is a function with domain equal to K' such that

$$q'(s, t) = q(s, t) \text{ if } (s, t) \in K' \text{ and } (s, t) \neq (z, u),$$

$$q'(z, u) = -q(u, z).$$

Next, we let $v' \in \mathscr{L}(K')$ be the branch voltage chain associated with the incidence function f' corresponding to the branch voltage chain $v \in \mathscr{L}(K)$ associated with the incidence function f. Since Ohm's Law, Kirchhoff's Voltage Law, and Kirchhoff's Current Law must be satisfied, we can assert that

$$v' \in \operatorname{rng} \delta',$$

$$Y'(v') - q' \in \operatorname{kernel} \Delta'.$$

We fix some branch $\{x, y\} \in S$ and we want to compare $v(f(\{x, y\}))$ with $v'(f'(\{x, y\}))$. In particular, we expect to show that if $\{x, y\} \neq \{u, z\}$ then $v(f(\{x, y\})) = v'(f'(\{x, y\}))$, while if $\{x, y\} = \{u, z\}$, then $v(f(\{x, y\})) = -v'(f'(x, y))$.

From Theorem 2 of 8.9, we know that

$$v(f(\{x, y\})) = J^{-1} r(\{x, y\}) \sum_{(s, t) \in K} q(s, t) \sum_{T \in \mathscr{T}} U(T) \, (B_T(\Phi^K_{f(\{x, y\})})) \, (s, t)$$

$$= J^{-1} r(\{x, y\}) \sum_{T \in \mathscr{T}} U(T) \sum_{(s, t) \in K} q(s, t) \, (B_T(\Phi^K_{f(\{x, y\})})) \, (s, t).$$

Similarly,

$$v'(f'(\{x, y\})) = J^{-1} r(\{x, y\}) \sum_{T \in \mathscr{T}} U(T) \sum_{(s, t) \in K'} q'(s, t) \, (B'_T(\Phi^{K'}_{f'(\{x, y\})})) \, (s, t).$$

Thus, to compare $v(f(\{x, y\}))$ with $v'(f'(\{x, y\}))$, we select some fixed $T \in \mathscr{T}$, and compare

$$\sum_{(s, t) \in K} q(s, t) \, (B_T(\Phi^K_{f(\{x, y\})})) \, (s, t)$$

with

$$\sum_{(s,\,t)\,\in\,K'} q'(s,t)\,(B'_T(\Phi^{K'}_{f'(\{x,\,y\})}))\,(s,t).$$

Now,

$$\sum_{(s,\,t)\,\in\,K} q(s,t)\,(B_T(\Phi^{K}_{f(\{x,\,y\})}))\,(s,t)$$

$$= \sum_{\{s,\,t\}\,\in\,S} q(f(\{s,t\}))\,(B_T(\Phi^{K}_{f(\{x,\,y\})}))\,(f(\{s,t\}))$$

and

$$\sum_{(s,\,t)\,\in\,K'} q'(s,t)\,(B'_T(\Phi^{K'}_{f'(\{x,\,y\})}))\,(s,t)$$

$$= \sum_{\{s,\,t\}\,\in\,S} q'(f'(\{s,t\}))\,(B'_T(\Phi^{K'}_{f'(\{x,\,y\})}))\,(f'(\{s,t\})).$$

Thus, we must compare

$$\sum_{\{s,\,t\}\,\in\,S} q(f(\{s,t\}))\,(B_T(\Phi^{K}_{f(\{x,\,y\})}))\,(f(\{s,t\}))$$

with

$$\sum_{\{s,\,t\}\,\in\,S} q'(f'(\{s,t\}))\,(B'_T(\Phi^{K'}_{f'(\{x,\,y\})}))\,(f'(\{s,t\})).$$

To do this, select some fixed $\{s,t\} \in S$.
We shall show that if $\{x,y\} \neq \{u,z\}$,

$$q(f(\{s,t\}))\,(B_T(\Phi^{K}_{f(\{x,\,y\})}))\,(f(\{s,t\}))$$

$$= q'(f'(\{s,t\}))\,(B'_T(\Phi^{K'}_{f'(\{x,\,y\})}))\,(f'(\{s,t\})),$$

while if $\{x,y\} = \{u,z\}$,

$$q(f(\{s,t\}))\,(B_T(\Phi^{K}_{f(\{x,\,y\})}))\,(f(\{s,t\}))$$

$$= -q'(f'(\{s,t\}))\,(B'_T(\Phi^{K'}_{f'(\{x,\,y\})}))\,(f'(\{s,t\})).$$

Now, by Lemma 3 of 8.7 we can assert that either $\{x,y\} \in T$ and $\{x,y\} = \{s,t\}$, or $\{x,y\} \in T$ and $\{x,y\} \neq \{s,t\}$ and $\{s,t\} \in S - T$ and $\{x,y\} \in \lambda(T,\{s,t\})$, for otherwise,

$$(B_T(\Phi^{K}_{f(\{x,\,y\})}))\,(f(\{s,t\})) = 0 = (B'_T(\Phi^{K'}_{f'(\{x,\,y\})}))\,(f'(\{s,t\})),$$

and the desired result is achieved.
The assumption that $\{x,y\} \in T$ guarantees that $B_T(\Phi^{K}_{f(\{x,\,y\})}) = d(T,\{x,y\})$ and $B'_T(\Phi^{K'}_{f'(\{x,\,y\})}) = d'(T,\{x,y\})$. Thus, we must show that if $\{x,y\} \neq \{u,z\}$,

$$q(f(\{s,t\}))\,(d(T,\{x,y\}))\,(f(\{s,t\})$$

$$= q'(f'(\{s,t\}))\,(d'(T,\{x,y\}))\,(f'(\{s,t\})),$$

while if $\{x,y\} = \{u,z\}$,

$$q\left(f(\{s,t\})\right)\left(d(T,\{x,y\})\right)\left(f(\{s,t\})\right)$$
$$= -q'\left(f'(\{s,t\})\right)\left(d'(T,\{x,y\})\right)\left(f'(\{s,t\})\right).$$

Two cases arise.

Case 1. $\{x,y\} \neq \{u,z\}$.

In this case, we must consider separately whether $\{s,t\} \neq \{u,z\}$ or $\{s,t\} = \{u,z\}$.

Thus, we suppose first that $\{s,t\} \neq \{u,z\}$, implying that $f'(\{s,t\}) = f(\{s,t\})$ and $q'(f'(\{s,t\})) = q(f(\{s,t\}))$.

Now, either $\{s,t\} = \{x,y\}$ or $\{s,t\} \neq \{x,y\}$.

If $\{s,t\} = \{x,y\}$, then

$$\left(d(T,\{x,y\})\right)\left(f(\{s,t\})\right) = -1 = \left(d'(T,\{x,y\})\right)\left(f(\{s,t\})\right),$$

leading to the desired result.

If $\{s,t\} \neq \{x,y\}$, then by our preceding assumption, $\{s,t\} \in S - T$ and $\{x,y\} \in \lambda(T,\{s,t\})$.

By Condition (3) of the definition of $\mu(T,\{x,y\})$ in 7.6, $\{s,t\} \in \mu(T,\{x,y\})$.

By Conditions (3)—(6) of the definition of $d(T,\{x,y\})$ in 7.6,

$$\left(d(T,\{x,y\})\right)\left(f(\{s,t\})\right) = \left(d'(T,\{x,y\})\right)\left(f(\{s,t\})\right),$$

leading to the desired result.

Suppose now that $\{s,t\} = \{u,z\}$, implying that $f(\{s,t\}) = (u,z)$ and $f'(\{s,t\}) = (z,u)$, and thus $q'(f'(s,t)) = -q(f(\{s,t\}))$.

But $\{s,t\} = \{u,z\}$ also implies that $\{s,t\} \neq \{x,y\}$, and by our preceding assumption, $\{s,t\} \in S - T$ and $\{x,y\} \in \lambda(T,\{s,t\})$.

By Condition (3) of the definition of $\mu(T,\{x,y\})$ in 7.6, $\{s,t\} \in \mu(T,\{x,y\})$.

By Conditions (3)—(6) of the definition of $d(T,\{x,y\})$ in 7.6,

$$\left(d(T,\{x,y\})\right)\left(f(\{s,t\})\right) = -d'(T,\{x,y\})\right)\left(f'(\{s,t\})\right).$$

Combining this fact with the assertion above that $q'(f'(\{s,t\})) = -q(f(\{s,t\}))$ gives the desired result.

Case 2. $\{x,y\} = \{u,z\}$.

Suppose first that $\{s,t\} \neq \{u,z\}$, implying that $f(\{s,t\}) = f'(\{s,t\})$ and $q(f(\{s,t\})) = q'(f'(\{s,t\}))$.

Also, $\{s,t\} \neq \{x,y\}$, so that by our preceding assumption, $\{s,t\} \in S - T$ and $\{x,y\} \in \lambda(T,\{s,t\})$.

By Condition (3) of the definition of $\mu(T,\{x,y\})$ in 7.6, $\{s,t\} \in \mu(T,\{x,y\})$.

By Conditions (3)–(6) of the definition of $d(T, \{x, y\})$ in 7.6,

$$(d(T, \{x, y\}))\, (f(\{s, t\})) = -\, (d'(T, \{x, y\}))\, (f(\{s, t\})),$$

leading to the desired result.

Finally, if $\{s, t\} = \{u, z\}$, then $f(\{s, t\}) = (u, z)$ and $f'(\{s, t\}) = (z, u)$, and thus, $q'(f'(\{s, t\})) = -\, q(f(\{s, t\}))$. But with $\{s, t\} = \{u, z\}$, $\{s, t\} = \{x, y\}$, and

$$(d(T, \{x, y\}))\, (f(\{s, t\})) = -1 = (d'(T, \{x, y\}))\, (f'(\{s, t\})),$$

which, combined with the assertion above that $q'(f'(\{s, t\})) = -\, q(f(\{s, t\}))$ gives the desired result.

Next, we let $i' \in \mathscr{L}(K')$ be the branch current chain associated with the incidence function f' corresponding to the branch current chain $i \in \mathscr{L}(K)$ associated with the incidence function f. Observe that both i and i' satisfy Ohm's Law and thus, $v = Z(i)$ and $v' = Z'(i')$. With our fixed branch $\{x, y\} \in S$ we expect to show that if $\{x, y\} \neq \{u, z\}$, then $i(f(\{x, y\})) = i'(f'(\{x, y\}))$, while if $\{x, y\} = \{u, z\}$, then $i(f(\{x, y\})) = -\, i'(f'(\{x, y\}))$.

But from the discussion of Ohm's Law in 6.4, we know that the fact that $v = Z(i)$ implies that

$$v(f(\{x, y\})) = r(\{x, y\})\, i(f(\{x, y\})),$$

while the fact that $v' = Z'(i')$ implies that

$$v'(f'(\{x, y\})) = r(\{x, y\})\, i'(f'(\{x, y\})).$$

Thus, from our previous facts relating $v(f(\{x, y\}))$ and $v'(f'(\{x, y\}))$, we obtain the desired results relating $i(f(\{x, y\}))$ and $i'(f'(\{x, y\}))$.

References

Below we list some references for supplementary reading. At the end of each such reference we list in parentheses the numbers of those Chapters of Chapters 1—8 of this book to which the reference applies.

ARSOVE, M. G.: A Note on the Network Postulates. Journal of Mathematics and Physics. **32**, 1953. (8)

BAYARD, M.: Theorie des Reseaux de Kirchhoff. La Revue d'Optique, Paris, 1954. (3, 6, 7, 8).

BERGE, C.: Theory of Graphs and its Applications, translated by A. DOIG. New York: John Wiley 1962. (1, 2).

BOURGIN, D. G.: Modern Algebraic Topology. New York: MacMillan 1963. (5).

CAIRNS, S. S.: Introductory Topology. New York: Ronald Press 1961. (5).

CAUER, W.: Synthesis of Linear Communication Networks. Translated by G. E. KNAUSENBERGER and J. N. WARFIELD. New York: McGraw-Hill 1958. (1, 2, 3, 6, 8).

CEDERBAUM, I.: Matrices All of Whose Elements and Subdeterminants are 1, —1, or 0. J. Math. Phys. **34**, 4, 1958. (3).

FOSTER, R. M.: Topological and Algebraic Considerations in Network Synthesis. Proc. Symposium on Modern Network Synthesis. New York: Polytechnic Institute of Brooklyn 1952. (1, 2, 3).

GUILLEMIN, E. A.: Introductory Circuit Theory. New York: John Wiley 1953. (2, 3, 6, 7).

HADLEY, G.: Linear Algebra. Reading, Mass.: Addison-Wesley 1961. (4).

HARARY, F.: Graph Theory and Electric Networks. IRE Trans. on Circuit Theory, special supplement, CT-6, May, 1959. (1, 2, 5, 6, 8).

HOCKING, J. G., and G. S. YOUNG: Topology. Reading, Mass.: Addison-Wesley 1961. (5).

HOFFMAN, K., and R. KUNZE: Linear Algebra. Englewood Cliffs, New Jersey: Prentice-Hall 1961. (4).

KAHN, P. J.: Introduction to Linear Algebra. New York: Harper and Row 1967. (4).

KIM, W. H., and R. T-W. CHIEN: Topological Analysis and Synthesis of Communication Networks. New York: Columbia University Press 1962. (1, 2, 3).

KIRCHHOFF, G.: On the Solution of the Equations Obtained from the Investigation of the Linear Distribution of Galvanic Currents. Translated by J. B. O'TOOLE. IRE Trans. on Circuit Theory, CT-5, March, 1958. (6, 8).

KÖNIG, D.: Theorie der Endlichen und Unendlichen Graphen. Leipzig: Akademische Verlagsgesellschaft 1936. (1, 2).

KURATOWSKI, C.: Sur le Probleme des Courbes Gauche en Topologie. Fundamenta Mathematicae **15**, 1930. (1).

MacWILLIAMS, F. J.: An Algebraic Proof of Kirchhoff's Network Theorem. Bell Telephone Laboratories, internal memorandum, Murray Hill, New Jersey, Oct. 22, 1958. (1, 2, 5, 6, 8).

NERING, E. D.: Linear Algebra and Matrix Theory. New York: John Wiley 1965. (4).

NERODE, A., and H. SHANK: An Algebraic Proof of Kirchhoff's Network Theorem. American Math. Monthly **68**, 3, March, 1961. (5, 6, 8).

—— Topological Network Theory. WADC Technical Report 57—424, Astia Document Number AD 155730, Wright Air Development Center, November, 1957. (1, 2, 5, 6, 7, 8).

ORE, O.: Graphs and Their Uses. New York: Random House 1963. (1, 2).

PONTRYAGIN, L. S.: Foundations of Combinatorial Topology. Translated by F. BAGEMIHL, W. KOMM, and W. SEIDEL. Rochester, New York: Graylock Press 1952. (5).

REZA, F. M.: Some Topological Considerations in Network Synthesis. IRE Trans. on Circuit Theory, CT-5, March, 1958. (1, 2, 6, 8).

— and S. SEELY: Modern Network Analysis. New York: McGraw-Hill 1959. (2, 6, 7).

SESHU, S., and N. BALABANIAN: Linear Network Analysis. New York: Wiley 1959. (2, 6, 7).

— and M. B. REED: Linear Graphs and Electrical Networks. Reading, Mass.: Addison-Wesley 1961. (1, 2, 3).

SHIELDS, P. C.: Linear Algebra. Reading, Mass.: Addison-Wesley 1964. (4).

TELLEGEN, B. D. H.: Theorie der Electrischen Netwerken. In ELIAS, G. J., and B. D. H. TELLEGEN: Theorie der Wechselstrommen, vol. 3. Groningen, Djakarta: P. Noordhoff N. V. 1952. (1, 2, 7, 8).

WEINBERG, L.: Kirchhoff's Third and Fourth Laws. IRE Trans. on Circuit Theory, CT-5, March, 1958. (1, 2, 3, 8).

— Network Analysis and Synthesis. New York: McGraw-Hill 1962. (1, 2, 3, 7, 8).

Index

Springer Tracts in Natural Philosophy